Developments in Mathematics

VOLUME 40

Series Editors:
Krishnaswami Alladi, *University of Florida, Gainesville, FL, USA*
Hershel M. Farkas, *Hebrew University of Jerusalem, Jerusalem, Israel*

More information about this series at http://www.springer.com/series/5834

Titu Andreescu • Dorin Andrica

Quadratic Diophantine Equations

Foreword by Preda Mihăilescu

Springer

Titu Andreescu
School of Natural Sciences
and Mathematics
University of Texas at Dallas
Richardson, TX, USA

Dorin Andrica
Faculty of Mathematics
and Computer Science
"Babeș-Bolyai" University
Cluj-Napoca, Romania

ISSN 1389-2177   ISSN 2197-795X (electronic)
Developments in Mathematics
ISBN 978-1-4939-3880-3   ISBN 978-0-387-54109-9 (eBook)
DOI 10.1007/978-0-387-54109-9

Mathematics Subject Classification (2010): 11B39, 11D09, 11D25, 11D41, 11D45, 11J70

Springer New York Heidelberg Dordrecht London

Springer Science+Business Media LLC New York is part of Springer Science+Business Media (www.
springer.com)

Foreword

When thinking of *quadratic Diophantine equations*, a flow of associations come to one's mind. Gauss's quadratic forms and their relation to class groups, fundamental units of real quadratic fields and Pell's equation, representations of integers as sums of two or four squares, or the apparently innocent equation $p = x^2 + ny^2$ and its relation to norm forms of imaginary quadratic fields and Kronecker's Jugendtraum—all these range from relatively simple subjects for introductory courses to elegant and involved topics of classical mathematics. But then one can go on: representation of integers by ternary quadratic forms, the equation $p = x^2 - ny^2$ and the *Manin's mid-life dream*, so one suddenly lands in the midst of highly interesting contemporary research topics in number theory. No, quadratic Diophantine equations must not be dull!

So, how much of it can be communicated to a large public without assuming involved prerequisites of algebra and number theory? This imposes some additional restrictions which make it unadvisable to approach topics beyond the general context of real quadratic fields and continuous fractions. The present book will succeed to convince the reader of the richness of topics that are available within this strict setting, as well as providing a fascinating and surprising list of applications of quadratic equations.

Already the first chapter offers a list of interesting and less well-known applications and connections of quadratic equations to various topics from, in part, unexpected areas of mathematics such as Hecke operators and Hecke groups, Einstein manifolds and geodesics, or projective geometry. The second and third chapters cover the basic knowledge on continued fractions, real quadratic fields, and Pell's equation; but they also offer the familiarized reader some interesting surprises of the kind that make this book attractive. For instance, the treatment of some special cases of the binary Thue equation $ax^2 - by^2 = 1$ and an investigation of the quadratic fields with units of negative norm, due to Stevenhagen.

On base of these prerequisites, the following chapters become more specialized: they cover a large variety of variants and equations reducible by not-so-obvious tricks to the Pell equation. Among them, a rich collection of representation problems of the kind $x^2 - ny^2 = N$, of representations which we mentioned being related to the real variant of Kronecker's Jugendtraum, but which can in these cases be solved by elementary methods.

We let the reader discover the unexpected applications and connections between Pell's equations and so many, surprising, problems, and decide which he or she finds most interesting and appealing. The book generously offers for all tastes.

Mathematisches Institut der Universität Göttingen   Preda Mihăilescu
Göttingen, Germany

Preface

Even though this domain seems to be classic, because of the last two decades of remarkable progress in computational technique, many open problems in Diophantine Analysis made a vigorous comeback with the hope that this increased power in computation will shade new light on them. The distinctive dynamic of Diophantine Analysis is expressively reflected in the well-known review journals: "Mathematical Reviews" (USA) and "Zentralblatt für Matematik" (now zbMATH) (Germany), the following AMS Subject Classification 2000 being designated to it: 11Dxx (Diophantine equations), 11D09 (Quadratic bilinear equations), 11A55 (Continued fractions), 11J70 (Continued fractions and generalizations), 11Y65 (Continued fraction), 11B37 (Recurrences), 11B39 (Fibonacci and Lucas numbers and polynomials and generalizations), 11R27 (Units and factorization). While in 1912 a first monograph [224] dedicated to this subject synthesized the important results known by then, a variety of papers and doctoral theses in Diophantine Analysis followed. Some of them are very recent, and we mention here just the monograph [32], published in 2003 and the syntheses [111] and [226], published in 2002.

The volume has seven chapters and moves gradually from the history and motivation of quadratic Diophantine equations to more complicated equations and their applications. The accessibility has also been taken into account, in the sense that some material could be understood even by nonspecialists in the field. In many sections the history of the advances and the problems that still remain open are also given, together with comments. Throughout this work, complete proofs are given and the sign $\square$ denotes the end of a proof.

Chapter 1, entitled "Why Quadratic Diophantine Equations?," describes the motivation of studying quadratic Diophantine equations. It contains ten short sections, each of them illustrating an important problem whose solution reduces to solving such an equation. Section 1.1 introduces the famous theorem of A. Thue [62] that reduces a Diophantine equation of the form $F(x,y) = m$ to a quadratic one, pointing out the importance of the quadratic case. Section 1.2 reviews Hilbert's tenth problem, which asks, in essence, for a finite algorithm to solve a general multivariable Diophantine equation. The proof of the impossibility of determining such an algorithm was found by Yu. Matiyasevich, who used the results of M.

Davis, H. Putnam, and J. Robinson. A crucial point in Matiyasevich's proof is the consideration of the solutions to the equations $x^2 - (a^2 - 1)y^2 = 1$, where $a \geq 2$. Section 1.3 presents the connection between Euler concordant forms and elliptic curves, the determination of the ranks of Mordell–Weil groups, and Frey curves, problems that lead to studying the solutions to two simultaneous Pell's equations. Section 1.4 makes the connection with the spectrum of the Laplace–Bertrami operator whose spectrum description involves solving a general Pell's type equation. In Section 1.5, a quadrature formula in the s-dimensional unit cube is considered, where the determination of parameters that minimize the remainder also reduces to solving some special Pell's type equations. The last five sections of this chapter briefly review: reduction algorithms of "threshold" phenomena in arbitrary lattices, the study of Einstein type Riemann metrics on homogeneous manifolds generated by the action of a semisimple Lie group, the problem of counting the number of autointersection points of the geodesics closed in the modular group $SL(2, \mathbb{Z})$, Hecke groups and their connection with continued fractions, as well as (m, n) type sets in the projective plane, all of these representing examples of problems from various areas of mathematics, whose solutions require solving Pell's type equations. Clearly, the list of such problems can be continued.

Chapter 2, "Continued Fractions, Diophantine Approximation and Quadratic Rings," presents two basic instruments of investigation in Diophantine Analysis. The theory of continued fractions plays an important part in pure Mathematics and has multiple applications. The main results of this theory, which are necessary in developing efficient algorithms in Diophantine Analysis, are given in Section 2.1. Among the references used, we mention here: [1, 46, 141, 159, 164, 183, 208]. The following aspects are introduced: the Euclid algorithm and its connection with continued fractions, the problem of uniqueness of a continued fraction developing, infinite continued fractions and their connection with irrational numbers, the approximation of irrational numbers by continued fractions, the problem of the best approximation and Hurwitz's Theorem, periodic continued fractions. The second important tool in studying problems in Diophantine Analysis is the theory of quadratic rings. The fundamental concepts, the units and norms defined in a natural way in this context are featured in Section 2.2. From the rich bibliography devoted to this subject, we mention [95, 171, 198].

Chapter 3, "Pell's Equation," is divided into six sections. The first is a comprehensive historical introduction to Pell's equation. Section 3.2 considers the problem of finding the general solution to Pell's equation by using elementary methods. The proof of the main theorem is based on our papers [13–15]. The main forms of writing the general solution are given: by recursive sequences, in matrix form, explicitly, etc. In Section 3.3, the general solution to Pell's equation is obtained by using the method of continued fractions. In Section 3.4, following the papers [171] and [95], the same problem is solved by using the theory of quadratic fields. Section 3.5 contains original contributions to the study of the more general equation $ax^2 - by^2 = 1$. Theorem 3.5.2 shows how one can determine the general solution to this equation, in case of solvability. The proof is the one given in our papers [13–15]. We note the fact that from our explicit form we found for the general solution to this equation (Remark 2) one gets immediately the result in [219], which is obtained there by a very complicated and unnatural method. The last section is dedicated to the negative Pell's equation $x^2 - Dy^2 = -1$. It is studied as a special case of the equation in the previous section and the central result concerning it is contained in Theorem 3.6.1. The formulas (3.6.3) and (3.6.4) give the general solution explicitly. The presentation follows again our papers [13–15]. Using now our paper [18], in Theorem 3.6.4 it is given a family of negative Pell's equations, solvable only for a single value of the positive integral parameter k. The section ends with the presentation of the current stage of the problem of solvability of the negative Pell's equation, an open problem that is far from being settled, and one of the most difficult in Diophantine analysis. In this respect, partial results, as well as recent conjectures, are mentioned.

The main goal of Chapter 4, "General Pell's Equation," is to present the general theory and major algorithms regarding the equation $x^2 - Dy^2 = N$. This chapter contains nine sections. In Section 4.1, the theory is exposed in a personal manner; the classes of solutions are defined, and Theorems 4.1.1, 4.1.2, and 4.1.3 give classical bounds for the fundamental solutions. These bounds were recently improved in [76] (L. Panaitopol, personal communication, December 2001). The section ends with our results concerning a problem proposed in [37], problem that we solve completely (see [16]). In addition, we present a final description of the set of all rational solutions to the Pell's equation, from which one can see clearly the complexity of the problem of determining this set explicitly. Also, an interesting result proved in [72] about the solvability of some general Pell's equation is mentioned. Section 4.2 contains results about the solvability of the general Pell's equation, and it is organized into five subsections dealing with the following aspects: Pell Decision Problem and the Square Polynomial Problem, the Legendre test, Legendre unsolvability tests, modulo n unsolvability tests, extended multiplication principle. Section 4.3 contains an algorithm for determining the fundamental solutions to the general Pell's equation, based on continued fractions. This algorithm is known as the LMM method. Numerical examples that probe the efficiency of the algorithm are also included. Section 4.4 deals with the problem of solving the general Pell's equation by using the PQa algorithm, derived from the theory of continued fractions, as well. A variant of the PQa algorithm for solving the negative Pell's equation, and the special Pell's equations $x^2 - Dy^2 = \pm 4$ are also given. Later in this section, the problem of the structure of the solutions to the general Pell's equation is taken on. By using the PQa algorithm, we study the problem of determining the fundamental solutions in the case $N < \sqrt{D}$ and then consider several numerical examples that illustrate how this algorithm works. In some of these examples, one compares the efficiency of the algorithm PQa versus the one of the LMM's method. Section 4.5 is dedicated to the study of the solvability and unsolvability of the equation $ax^2 - by^2 = c$. All results here belong to us and are based on our papers [13, 14, 17]. Two general methods for solving the above-mentioned equation are presented, and a complete answer to the problem posed in the recent paper [114] is given. An original point of view is contained in Theorem 4.5.2, where the solvability of this equation is linked to the solvability of two other quadratic equations. In Theorem 4.5.3, a large class of solvable equations $ax^2 - by^2 = c$ is given. Section 4.6 deals with the problem of solving the general Pell's equation by using the theory of quadratic rings. One obtains a different algorithm for solving the general Pell's equation than the one described in Section 4.7. The main goal of Section 4.8 is to discuss the more general equation $ax^2 + bxy + cy^2 = N$. Recently, this equation captured the attention of mathematicians (see, for example, the doctoral thesis [157]). The last section in this chapter is dedicated to the connection between the Thue's theorem (Theorem 4.9.1) and the equations $x^2 - Dy^2 = \pm N$. We discuss here the equation of this form with $D = 2, 3, 5, 6, 7$, the working method being directly obtained from the Thue's theorem.

Chapter 5 is called "Equations Reducible to Pell's Type Equations." In Section 5.1, we present the equations $x^2 - kxy^2 + y^4 = 1$ and $x^2 - kxy^2 + y^4 = 4$, the main results being contained in Theorems 5.1.1, 5.1.7, and 5.1.9. Section 5.2 is dedicated to the equation $x^{2n} - Dy^2 = 1$ and the central results are given in Theorems 5.2.3 and 5.2.4. Two special equations that finally lead to Pell's type equations are studied in Sections 5.3 and 5.4. Our point of view contained in section 5.5 encompasses in a unitary class several equations dispersed in the literature, for instance the equations in the titles of subsections 5.5.1, 5.5.2, 5.5.3, and 5.5.4. Section 5.6 points out other quadratic equations with infinitely many integral solutions. In subsection 5.6.1, we rely on our paper [12]. The main result concerning the equation $x^2 + axy + y^2 = 1$ is given in Theorem 5.6.1. In subsection 5.6.2, we study the equation (5.6.8), our results correcting the ones in [173]. Other interesting equations of this type, both solvable, are (5.6.13) and (5.6.14). The equation (5.6.15) is studied in our paper [8], where five distinct infinite families of solutions are displayed. Based on our paper [9], we find nine different infinite families of positive integral solutions to the equation (5.6.18). The main idea of the paper [9] was used in [79] to generate six infinite families of positive integral solutions to the equation (5.6.24). By using a result in our paper [10], in the last subsection of Section 5.6 we prove that equation (5.6.24) has in certain conditions infinitely many integral solutions.

In Chapter 6, "Diophantine Representations of Some Sequences," we study a first class of applications of some theoretical results presented in the previous chapters. In Section 6.1, we define the concept of Diophantine r-representability, making the connection with the papers [27, 28] and the doctoral Dissertation [47]. From Theorem 6.1.1, in Section 6.2 we then obtain as special cases some properties concerning Fibonacci's, Lucas', and Pell's sequences, given in (6.2.1), (6.2.2), and (6.2.3). In Section 6.3, we reconsider in an original manner the study of the equations $x^2 + axy + y^2 = \pm 1$. The central result is contained in Theorem 6.3.1 and its proof is based on solving the special Pell's equation $u^2 - 5v^2 = -4$. All results in Section 6.4, concerning the equation (6.4.1) and its connection with the Diophantine representation of the Fibonacci, Lucas, and Pell sequences are original. The method we employ is different, more natural, and simpler than the one in the doctoral dissertation [47]. Section 6.5 deals with the Diophantine representation of the generalized Lucas sequences, defined by us in (6.5.1). The main results, given in Theorems 6.5.1 and 6.5.2, generalize the ones in Section 6.4. Some special cases are considered in the papers [96, 97, 134] and [61], while a particular definition of Lucas sequences is given in [91]. Important results spelling out the conditions in which the solutions to the equation (6.5.8) are linear combinations with rational coefficients of Fibonacci and Lucas classic sequences are contained in Theorems 6.5.4 and 6.5.5. These results belong to us as well and appear in our papers [19] and [20].

The last chapter, "Other Applications," contains five sections. Based on our papers [13] and [14], we extend the results in [122] and [123] concerning the conditions in which the numbers $an + b$ and $cn + d$ are simultaneously perfect squares for infinitely many values of the positive integer n. The main result is given in Theorem 7.1.1 and is based on Theorem 4.5.3. Several special cases appear in [25, 199] and [40]. Section 7.2 is dedicated to the study of some special properties of the triangular numbers. In Theorem 7.2.1, one determines all such numbers that are perfect squares, while in Theorem 7.2.2 one studies an equation solvable in the set of triangular numbers. The fact that the asymptotic density of the triangular numbers is equal to 0 is proved in what follows. Theorem 7.2.3 specifies that equation (7.2.9) is solvable for infinitely many triples (m, n, p) of positive integers and unsolvable for infinitely many triples (m, n, p) of positive integers. In [170], it is shown (Theorem 7.2.4) that any positive rational number r with $\sqrt{r} \notin Q$ can be written as a ratio of two triangular numbers. The proof of this theorem uses in an essential way our result contained in Theorem 4.5.3. Also in this section, we solve completely the problem of finding all triangular numbers T_m, T_n such that T_m/T_n is the square of a positive integer. Our approach generalizes the method given in [139] and [101]. In Section 7.3, we study some properties of the polygonal numbers that generalize the ones concerning triangular numbers (Theorem 7.3.1). In Section 7.3 we present some important results about powerful numbers. Recall that a positive integer r is a powerful number if p^2 divides r whenever the prime p divides r. We prove the theorem of asymptotic behavior of the function $k(x)$, where $k(x)$ is the number of powerful numbers less than or equal to x and, finally, we present results concerning consecutive powerful numbers and the possible distances between two powerful numbers. These results are also based on the general theory of quadratic equations.

In the last section, we present the solution to an open problem involving matrices in the ring $M_2(\mathbb{Z})$. The approach uses the properties of some quadratic Diophantine equations, and it was given in the recent paper [29].

Acknowledgements We feel deeply honored and grateful to Professor Preda Mihăilescu for his Foreword.

Special thanks are given to Dr. Diana Savin and to the anonymous referee for the pertinent suggestions and remarks that directly contributed to the improvement of the manuscript.

Finally, we would like to thank Dr. John P. Robertson for his useful suggestions.

Richardson, TX, USA   Titu Andreescu
Cluj-Napoca, Romania   Dorin Andrica
March 2015

Contents

List of Symbols

$\langle a_0, a_1, a_2, \ldots \rangle$	Infinite simple continued fraction
$\langle \overline{a_0, a_1, \ldots, a_n} \rangle$	Purely periodic continued fraction
$[x]$	Greatest integer $\leq x$
$\square$	End of proof of theorems, lemmas and corollaries
$\mathbb{Q}(\sqrt{D})$	Quadratic field
$\mu = m + n\sqrt{D}$	Quadratic surd
$\overline{\mu} = m - n\sqrt{D}$	Conjugate of μ
$N(\mu) = m^2 - Dn^2$	Norm of μ
$a\|b$	a divides b
$a \equiv b \pmod{m}$	a is congruent to b modulo m
$a \not\equiv b \pmod{m}$	a is not congruent to b modulo m
$\mathbb{Z}$	Set of integers
$\mathbb{Q}$	Set of rationals
$ax^2 + bxy + cy^2$ $+ dx + ey + f = 0$	Quadratic equation in x, y
$x^2 - Dy^2 = 1$	Pell's equation
$x^2 - Dy^2 = -1$	Negative Pell's equation
$x^2 - Dy^2 = N$	General Pell's equation
$x^2 - Dy^2 = 4$	Special Pell's equation
$x^2 - Dy^2 = -4$	Negative Pell's equation
$|x|$	Absolute value of real number x
$A = [a_{ij}]_{1 \leq i,j \leq n}$ $= (a_{ij})_{1 \leq i,j \leq n}$	Square matrix A of size n by n
I_n	Identity matrix of size n by n
$\det A$	Determinant of square matrix A
$\ln x$	Natural logarithm of x

E_n	n^{th} Euler number
$\|a\|$	Euclidean norm of $a \in \mathbb{R}^2$
$\left(\dfrac{a}{p}\right)$	Legendre symbol
$\dbinom{n}{k}$	Binomial coefficient
F_n	n^{th} Fibonacci number
L_n	n^{th} Lucas number
P_n	n^{th} Pell number
$k(D,N)$	Number of class of solutions of $x^2 - Dy^2 = N$
$\mathcal{K}(D,N)$	Set of fundamental solutions of classes of $x^2 - Dy^2 = N$
PDP	Pell Decision Problem
PSP	Pell Search Problem
SPDP	Square Polynomial Decision Problem
B_n	n^{th} Bernoulli number
$\gcd(x,y)$	Greatest common divisor of x, y
T_n	n^{th} triangular number
P_k^n	k^{th} n-gonal number
$k(x)$	Number of powerful numbers not exceeding x
ζ	Riemann zeta function
$M_2(\mathbb{Z})$	Ring of square matrices with integer entrees

Chapter 1
Why Quadratic Diophantine Equations?

In order to motivate the study of quadratic type equations, in this chapter we present several problems from various mathematical disciplines leading to such equations. The diversity of the arguments to follow underlines the importance of this subject.

1.1 Thue's Theorem

Since ancient times mathematicians tried to solve equations over the integers. Pythagoras for instance described all integers as side lengths of rectangular triangles. After Diophantus from Alexandria such equations are called *Diophantine equations*. Since that time, many mathematicians worked on this topic, such as Fermat, Euler, Kummer, Siegel, and Wiles.

In 1909, A. Thue (see [62]) proved the following important theorem:

Let $f = a_n z^n + a_{n-1} z^{n-1} + \cdots + a_1 z + a_0$ be an irreducible polynomial of degree ≥ 3 with integral coefficients. Consider the corresponding homogeneous polynomial

$$F(x, y) = a_n x^n + a_{n-1} x^{n-1} y + \cdots + a_1 x y^{n-1} + a_0 y^n.$$

If m is a nonzero integer, then the equation

$$F(x, y) = m$$

has either no solution or only a finite number of solutions in integers.

This result is in contrast to the situation when the degree of F is $n = 2$. In this case, if $F(x, y) = x^2 - Dy^2$, where D is a nonsquare positive integer, then for all nonzero integers m, the general Pell's equation

$$x^2 - Dy^2 = m$$

has either no solution or it has infinitely many integral solutions.

© Springer Science+Business Media New York 2015
T. Andreescu, D. Andrica, *Quadratic Diophantine Equations*,
Developments in Mathematics 40, DOI 10.1007/978-0-387-54109-9_1

1.2 Hilbert's Tenth Problem

In 1900, at the International Congress of Mathematicians in Paris, David Hilbert, looking forward to the coming century, proposed 23 problems with which twentieth century mathematicians would have to contend. The tenth on the list, commonly simply termed "Hilbert's Tenth Problem," called for a general method to determine the solvability or unsolvability in integers of Diophantine equations. With our current knowledge of the unsolvability of this problem, it would be interesting to know why Hilbert felt it had a positive solution. Did it look like the Gaussian theory could extend indefinitely to more and more variables and higher and higher degrees? Did he think one could cut through all the details and give an abstract proof, in the way he finished off the theory of invariants? Or was it just a manifestation of his faith in the mathematician's ability to solve all problems he posed for himself—a faith on which he was quite explicit? The actual statement of Hilbert's Tenth Problem is rather brief and uninformative:

Given a Diophantine equation with any number of unknown quantities and with integral numerical coefficients, devise a process according to which it can be determined by a finite number of operations whether the equation is solvable in integers.

To solve the Diophantine equation

$$f(x_1, x_2, \ldots, x_n) = 0,$$

where $f \in \mathbb{Q}[X_1, X_2, \ldots, X_n]$, amounts to determine the integer points on the corresponding hypersurface of the affine space. Hilbert's tenth problem is to give an algorithm which tells whether such a given Diophantine equation has a solution or not.

The final answers to Hilbert original tenth problem were given in 1970 by Yu. Matiyasevich, after the works of M. Davis, H. Putnam, and J. Robinson. This was the culminating stage of a rich and beautiful theory (see [59, 124, 130, 131], and [132]). The solution is negative: there is no hope nowadays to achieve a complete theory of the subject. But one may still hope that there is a positive answer if one restricts Hilbert's initial question to equations in few variables, say $n = 2$, which amounts to considering integer points on a plane curve. In this case deep results have been achieved during the twentieth century and many results are known, but much more remains unveiled.

The logical assault on Hilbert's Tenth Problem began around 1950, the first tentative papers appearing in the ensuing decade, the first major breakthrough appearing in print in 1961, and the ultimate solution being published in 1970. The first contributions were made by Julia Robinson and Martin Davis.

Robinson defined a relation R on natural numbers to be *Diophantine* if it could be written in the form

$$R(x_0, \ldots, x_{n-1}) : \exists y_0 \ldots y_{m-1} P(x_0, \ldots, x_{n-1}, y_0, \ldots, y_{m-1}) = 0,$$

where P is a polynomial with integral coefficients and $y_0, \ldots, y_{m-1}$ range over *natural* numbers. (Logicians prefer their Diophantine equations to have nonnegative integral solutions, an inessential reformulation of the usual Fermatian Diophantine problem.) Finding she could not exhibit many demonstrably Diophantine relations, she allowed exponentiation to enter into P to form *exponential Diophantine* relations. She was able to show several interesting relations to be exponential Diophantine, and she reduced the general problem of showing all exponential Diophantine relations to be Diophantine to that of showing any relation of roughly exponential growth to have a Diophantine graph. In this reduction, she used the sequence of solutions to the special Pell's equations

$$x^2 - (a^2 - 1)y^2 = 1, \quad a \geq 2.$$

Davis took a more logical approach. The theory of algorithms recognizes two basic types of sets of natural numbers, namely: *recursive* sets, for which an algorithm determining membership exists, and *recursively enumerable* sets, for which an algorithmic enumeration exists. There are recursively enumerable sets which are not recursive. If every recursively enumerable set could be shown to be Diophantine, then Hilbert's Tenth Problem would have no effective solution. The techniques Gödel developed in proving his famous Incompleteness Theorems readily show that every recursively enumerable set can be written in the form

$$\exists y_0 Q_1 y_1 \ldots Q_{m-1} y_{m-1} P(x, y_0, \ldots, y_{m-1}) = 0,$$

where each Q_i is either an existential quantifier or a *bounded* universal quantifier, i.e., a quantifier of the form $\forall y_i \leq y_0$. Davis simplified this representation to the *Davis Normal Form*

$$\exists y \forall z \leq y \exists w_0 \ldots w_{m-1} \leq y P(x, y, z, w_0, \ldots, w_{m-1}) = 0.$$

Within a few years, Robinson's husband Raphael showed one could take $m = 4$.

Towards the end of the 1950s, Hilary Putnam joined Davis. Together they proved—modulo the unproved assumption of the existence of arbitrarily long arithmetic progressions of prime numbers—the unsolvability of the exponential Diophantine problem over the natural numbers. With Julia Robinson's help, the unproven conjecture was bypassed. Together, Davis, Putnam, and Robinson applied Robinson's exponential Diophantine relations to eliminate the single bounded universal quantifier from the Davis Normal Form. Their proof was published in 1961.

With the Davis–Putnam–Robinson Theorem, Robinson's reduction of the problem of representing exponential Diophantine relations as Diophantine relations to the special problem of giving a Diophantine representation of a single relation of roughly exponential growth assumed a greater importance. The 1960s saw no progress in the construction of such a relation, merely a profusion of further reductions based on it; for example, on the eve of the final solution, Robinson showed it sufficient to prove the Diophantine nature of any infinite set of prime numbers. In March 1970 the world of logic learned that the then twenty-two-year-old Yuri Matiyasevich has shown the relation

$$y = F_{2x}$$

to be Diophantine, where $F_0, F_1, \ldots$ is the Fibonacci sequence. Very quickly, a number of researchers adapted Matiyasevich's proof to give a direct proof of the Diophantiness of the sequences of solutions to the special Pell's equations

$$x^2 - (a^2 - 1)y^2 = 1, \quad a \geq 2,$$

cited earlier.

Interestingly, the corresponding problems when real solutions are looked for has a positive answer. The decision problem over the reals: *is there an algorithm deciding the existence of real solutions to a set of polynomial equation with integer coefficients?* was solved with a yes answer by Tarski [211] and Seidenberg [197].

1.3   Euler's Concordant Forms

In 1780, Euler asked for a classification of those pairs of distinct nonzero integers M and N for which there are integer solutions (x, y, z, t) with $xy \neq 0$ to

$$x^2 + My^2 = t^2 \quad \text{and} \quad x^2 + Ny^2 = z^2.$$

This is known as Euler's concordant forms problem. When $M = -N$, Euler's problem is the celebrated congruent number problem to which Tunnell gave a conditional solution using the theory of elliptic curves and modular forms. More precisely, let $E_{M,N}$ be the elliptic curve

$$y^2 = x(x + M)(x + N).$$

If the group of $\mathbb{Q}$-rational points of $E_{M,N}$ has positive rank, then there are infinitely many primitive integer solutions to Euler's concordant forms problem. But if the rank is zero, there is a solution if and only if the $\mathbb{Q}$-torsion subgroup of rational points of $E_{M,N}$ is $\mathbb{Z}_2 \times \mathbb{Z}_8$ or $\mathbb{Z}_2 \times \mathbb{Z}_6$. All the curves $E_{M,N}$ having such torsion subgroups are classified as follows. The torsion subgroup of $E_{M,N}(\mathbb{Q})$ is $\mathbb{Z}_2 \times \mathbb{Z}_8$ if there is a nonzero integer d such that $(M, N) = (d^2u^4, d^2v^4)$ or $(-d^2v^4, d^2(u^4 - v^4))$ or $(d^2(u^4 - v^4), -d^2v^4)$, where (u, v, w) is a Pythagorean triple. The torsion subgroup of $E_{M,N}(\mathbb{Q})$ is $\mathbb{Z}_2 \times \mathbb{Z}_6$ if there exist integers a and b such that $\dfrac{a}{b} \notin \left\{-2, -1, -\dfrac{1}{2}, 0, 1\right\}$ and $(M, N) = (a^4 + 2a^3b, 2a^3b + b^4)$. Thus Euler's problem is reduced to a question of $\mathbb{Q}$-ranks of the Mordell–Weil groups of Frey curves. In proving the above results (see [166]) the study of the simultaneous Pell's equations

$$a^2 - Mb^2 = 1 \quad \text{and} \quad c^2 - Nb^2 = 1$$

played an important role.

1.4 Trace of Hecke Operators for Maass Forms

The trace of the Hecke operator $T(n)$ acting on a Hilbert space of functions spanned by the eigenfunctions of the Laplace–Beltrami operator Δ with a positive eigenvalue can be viewed as an analogue of Eichler–Selberg trace formula for nonholomorphic cusp forms of weight zero. For Re $\sigma > 1$, let

$$L_n(\sigma) = \sum_{d \in \Omega} \sum_u \frac{h_d \ln \varepsilon_d}{(du^2)^\sigma},$$

where the summation on u is taken over all the positive integers u which together with t are the integral solution of the equation $t^2 - du^2 = 4n$, h_d is the class number of indefinite rational quadratic forms with discriminant d, and $\varepsilon_d = \dfrac{1}{2}(u_0 + v_0\sqrt{d})$, with (u_0, v_0) being the fundamental solution to the general Pell's equation

$$u^2 - dv^2 = 4.$$

Here Ω is the set of all positive integers d such that $d \equiv 0$ or $1 \pmod 4$ and d is not a perfect square. For more details we refer to [113].

1.5 Diophantine Approximation and Numerical Integration

Consider the quadrature formula over the s-dimensional unit cube of the form

$$\int_0^1 \cdots \int_0^1 f(x_1, \ldots, x_s)dx_1 \ldots dx_s = q^{-1} \sum_{t=1}^q f\left(\frac{a_1 t}{q}, \ldots, \frac{a_s t}{q}\right) + R,$$

where q and $a_1, \ldots, a_s$ are positive integers and f is supposed periodic of period 1 in each variable. Choose q to be a prime and seek to determine $a_1, \ldots a_s$ so as to minimize R for a class of functions f whose multiple Fourier coefficients are small.

One of the main methods is based on units of algebraic fields (see [94]). In this method the field is $R(\sqrt{p_1}, \ldots, \sqrt{p_t})$, where $p_1, \ldots, p_t$ are distinct primes and the units are the fundamental solutions to the general Pell's equations

$$x^2 - Dy^2 = \pm 4,$$

where D runs over the $2^t - 1$ proper divisors of $p_1 p_2 \ldots p_t$.

1.6   Threshold Phenomena in Random Lattices and Reduction Algorithms

By a lattice is meant here the set of all linear combinations of a finite collection of vectors in $\mathbb{R}^n$ taken with integer coefficients,

$$\mathcal{L} = \mathbb{Z}e_1 \oplus \cdots \oplus \mathbb{Z}e_p.$$

One may think of a lattice as a regular arrangement of points in space, somewhat like atoms composing a crystal in $\mathbb{R}^3$. Given the generating family (e_j), there is great interest in finding a "good" basis of the lattice. By this is meant a basis that is "almost" orthogonal and is formed with vectors of "small" length. The process of constructing a "good" basis from a skewed one is referred to as lattice reduction.

Lattice reduction is of structural interest in various branches of mathematics. For instance, reduction in dimension 2 is completely solved by a method due to Gauss. This entails a complete classification of binary quadratic forms with integer coefficients, a fact that has numerous implications in the analysis of quadratic irrationals and in the representation of integers by quadratic forms, for example Pell's equation

$$x^2 - dy^2 = 1.$$

1.7   Standard Homogeneous Einstein Manifolds and Diophantine Equations

If $M = G/H$ is a homogeneous manifold and G is a semisimple Lie group, then there is a standard Riemannian metric g on M given by restricting the Killing form. An interesting problem is to study when g is an Einstein metric. In many cases the Einstein equations reduce to a series of integer constraints. These Diophantine equations in certain special cases often reduce to variants of Pell's equation. Such a case is as follows. Suppose K, L and R are Lie groups. Let $G = K^r \times R$, for some integer r, and let $H = K \times L$. Suppose that H is embedded in G via (Δ, π), where $\Delta : K \to K^r$ is the diagonal and $\pi : K \times L \to R$ is some representation. There are limited number of possibilities for K, L and R. One of these is $K = SO(n)$, $L = SO(m)$ and $R = SO(n + 1)$. The Einstein equations reduce to some Pell's equations in n, m and r and these have infinitely many solutions (see [156] for details).

1.8   Computing Self-Intersections of Closed Geodesics

Let Γ be a subgroup of finite index in the modular group $\Gamma(1) = SL(2, \mathbb{Z})$. An interesting problem is to compute the self-intersection number of a closed geodesic on H/Γ, where H is the hyperbolic plane.

Suppose the geodesic is given as the axis γ of $A \in \Gamma(1)$. Since by assumption $[\Gamma(1) : \Gamma] < \infty$, one can find a least $n \in \mathbb{Z}_+$, with $A^n \in \Gamma$. Fix the standard fundamental region $R = \left\{ z \in H : |z| \geq 1, |\mathrm{Re}\, z| \leq \frac{1}{2} \right\}$ for $\Gamma(1)$ and let $\mathcal{T}$ be the tessellation of H by images of R under $\Gamma(1)$. As γ cuts $\mathcal{T}$, it is divided into segments. Translating back to R by appropriate products of the generators $\begin{pmatrix} 1 & 1 \\ 0 & 1 \end{pmatrix}$ and $\begin{pmatrix} 0 & -1 \\ 1 & 0 \end{pmatrix}$, determined by the order in which γ cuts sides of $\mathcal{T}$, one can obtain a finite family of segments in R whose union projects to cover a fundamental period of A^n disjointly (except that points on ∂R will be covered twice). The algorithm consists in computing the endpoints on R of all these translates of γ (under $\Gamma(1)$) and testing them in pairs for intersections in $\Gamma(1)$ and then in Γ.

It is possible to write down the primitive hyperbolic in $\Gamma(1)$ whose axis joins two conjugate quadratic numbers, using minimal solutions of suitable *Pell's equations* (see [110]).

1.9   Hecke Groups and Continued Fractions

The Hecke groups

$$
G_q = \left\langle \begin{pmatrix} 1 & \lambda_q \\ 0 & 1 \end{pmatrix}, \begin{pmatrix} 0 & -1 \\ 1 & 0 \end{pmatrix} \right\rangle ; \quad \lambda_q = 2 \cos \frac{\pi}{q}, \quad q \geq 3
$$

are Fuchsian groups of the first kind. The λ-continued fractions (λF) can be used to study the geodesics on the modular surfaces determined by G_q. The period of the λF for periodic $\sqrt{D}/C$ has nearly the form of the classical case. The solutions to *Pell's equation* in quadratic extensions of $\mathbb{Q}(\lambda_q)$ as well as the Legendre's constant of Diophantine approximation for G_q, i.e., γ_q such that $\left| \alpha - \frac{P}{Q} \right| < \frac{\gamma_q}{Q^2}$ implies that $\frac{P}{Q}$ of "reduced finite λF form" is a convergent of real $\alpha \notin G_q(\infty)$, play an important role in proving the above result. For details we refer to [186].

1.10   Sets of Type (m, n) in Projective Planes

A set of type (m, n) in a projective plane is a set of points such that each line intersects it in either m or n points. Numerical conditions for the existence of such sets in planes of finite order q can be given. In particular, for $m = 1$ and $n \geq 4$, it is shown that q is of the form $q = (n - 1)P_s(n)$, where $P_s(n)$ satisfies the recurrence relation $P_0(n) = 0$, $P_1(n) = 1$, and $P_s(n) = (n - 2)P_{s-1}(n) - P_{s-2}(n) + 1$. The proof consists in solving the quadratic Diophantine equation in two variables $x^2 - x - (n - 2)xy + y^2 - y = 0$ which is related to a *general Pell's equation* (see [209] for details).

Chapter 2
Continued Fractions, Diophantine Approximation, and Quadratic Rings

The main goal of this chapter is to lay out basic concepts needed in our study in Diophantine Analysis. The first section contains fundamental results pertaining to continued fractions, some without proofs. The Theory of Continued Fractions is not new but it plays a growing role in contemporary mathematics.

Continued fractions have fascinated mankind for centuries if not millennia. The timeless construction of a rectangle obeying the "divine proportion" (the term is in fact from the Renaissance) and the "self-similarity" properties that go along with it are nothing but geometric counterparts of the continued fraction expansion of the golden ratio,

$$\phi \equiv \frac{1+\sqrt{5}}{2} = \cfrac{1}{1+\cfrac{1}{1+\cfrac{1}{1+\dots}}}.$$

Geometry was developed in India from the rules for the construction of altars. The Sulva Sutra (a part of the Kalpa Sutra hypothesized to have been written around 800 BC) provides a rule for doubling an area that corresponds to the near-equality:

$$\sqrt{2} = 1 + \frac{1}{3} + \frac{1}{3 \times 4} - \frac{1}{3 \times 4 \times 34} \quad \text{(correct to } 2 \cdot 10^{-6}).$$

The third and fourth partial sums namely $\frac{17}{12}$ and $\frac{577}{408}$ are respectively the fourth and eight convergents to $\sqrt{2}$.

Accordingly, in the classical Greek world, there is evidence of knowledge of the continued fraction for $\sqrt{2}$ which appears in the works of Theon of Smyrna (discussed in Fowler's reconstruction [74] and in [215]) and possibly of Plato in *Theatetus*, see [49]. As every student knows, Euclid's algorithm is a continued fraction expansion algorithm in disguise, and Archimedes' Cattle Problem (circa 250 BC)

© Springer Science+Business Media New York 2015
T. Andreescu, D. Andrica, *Quadratic Diophantine Equations*,
Developments in Mathematics 40, DOI 10.1007/978-0-387-54109-9_2

most probably presupposes on the part of its author some amount of understanding of quadratic irrationals, Pell's equation, and continued fractions; see [215] for a discussion.

The continued fraction convergent $\pi \approx \dfrac{355}{113}$ was known to Twu Ch'ung Chi, born in Fan-yang, China in 430 AD. More recently, the Swiss mathematician Lambert proved the 2,000 year conjecture (it already appears in Aristotle) that π is irrational, this thanks to the continued fraction expansion of the tangent function,

$$\tan z = \cfrac{z}{1 - \cfrac{z^2}{3 - \cfrac{z^2}{5 - \dots}}},$$

and Apéry in 1979 gave in "a proof that Euler missed" [176] nonstandard expansions like

$$\zeta(3) = \sum_{n=1}^{\infty} \frac{1}{n^3} = 1 + \cfrac{1}{2 \cdot 2 + \cfrac{1^3}{1 + \cfrac{1^3}{2 \cdot 6 + \cfrac{2^3}{1 + \cfrac{2^3}{2 \cdot 10 + \cfrac{3^3}{1 + \dots}}}}}}$$

from which the irrationality of $\zeta(3)$ eventually derives.

The standard method to prove the irrationality of e^x for nonzero rational x is by obtaining a rational approximation using the differential and integral properties of e^x and the differential properties of $x^n(1-x)^n/n!$, see [88]. Recently, a simple proof by using the theory of continued fractions was given in [154].

The principal references used in this section are [1, 46, 66, 141, 159, 164, 183, 184, 208].

The Section 2.2 presents key results regarding quadratic rings, their units and norms defined in a natural way. Important references for this section are [95, 171, 198].

2.1   Simple Continued Fractions

2.1.1   The Euclidean Algorithm

Given any rational fraction u_0/u_1, in lowest terms so that $\gcd(u_0, u_1) = 1$ and $u_1 > 0$, we apply the Euclidean algorithm (see [21]) to get successively

$$u_0 = u_1 a_0 + u_2, \qquad 0 < u_2 < u_1$$
$$u_1 = u_2 a_1 + u_3, \qquad 0 < u_3 < u_2$$
$$u_2 = u_3 a_2 + u_4, \qquad 0 < u_4 < u_3 \tag{2.1.1}$$
$$\cdots$$
$$u_{j-1} = u_j a_{j-1} + u_{j+1}, \, 0 < u_{j+1} < u_j$$
$$u_j = u_{j+1} a_j.$$

If we write ξ_i in place of u_i/u_{i+1} for all values of i with $0 \le i \le j$, then equations (2.1.1) become

$$\xi_i = a_i + \frac{1}{\xi_{i+1}}, \quad 0 \le i \le j - 1; \quad \xi_j = a_j. \tag{2.1.2}$$

If we take the first two of these equations, those for which $i = 0$ and $i = 1$, and eliminate ξ_1, we get

$$\xi_0 = a_0 + \cfrac{1}{a_1 + \cfrac{1}{\xi_2}}.$$

In this result we replace ξ_2 by its value from (2.1.2), and then we continue with replacement of $\xi_3, \xi_4, \ldots$, to get

$$\frac{u_0}{u_1} = \xi_0 = a_0 + \cfrac{1}{a_1 + \cfrac{}{\ddots \; + \cfrac{1}{a_{j-1} + \cfrac{1}{a_j}}}} \tag{2.1.3}$$

This is a *continued fraction expansion* of ξ_0, or of u_0/u_1. The integers a_i are called the *partial quotients* since they are the quotients in the repeated application of the division algorithm in equations (2.1.1). We presumed that the rational fraction u_0/u_1 had positive denominator u_1, but we cannot make a similar assumption about u_0. Hence a_0 may be positive, negative, or zero. However, since $0 < u_2 < u_1$, we note that the quotient a_1 is positive, and similarly the subsequent quotients $a_2, a_3, \ldots, a_j$ are positive integers. In case $j \ge 1$, that is if the set (2.1.1) contains more than one equation, then $a_j = u_j/u_{j+1}$ and $0 < u_{j+1} < u_j$ imply that $a_j > 1$.

We will use the notation $\langle a_0, a_1, \ldots, a_j \rangle$ to designate the continued fraction in (2.1.3). In general, if $x_0, x_1, \ldots, x_j$ are any real numbers, all positive except perhaps x_0, then we will write

$$\langle x_0, x_1, \ldots, x_j \rangle = x_0 + \cfrac{1}{x_1 + \cfrac{\ddots}{}}$$

$$+ \cfrac{1}{x_{j-1} + \cfrac{1}{x_j}}.$$

Such a finite continued fraction is said to be *simple* if all the x_i are integers. The following notations are often used to simplify the writing:

$$\langle x_0, x_1, \ldots, x_j \rangle = x_0 + \cfrac{1}{\langle x_1, \ldots, x_j \rangle}$$

$$= \left\langle x_0, x_1, \ldots, x_{j-2}, x_{j-1} + \frac{1}{x_j} \right\rangle.$$

The symbol $[x_0, x_1, \ldots, x_j]$ is sometimes used to represent a continued fraction. We use the notation $\langle x_0, x_1, \ldots, x_j \rangle$ to avoid confusion with the least common multiple and the greatest integer.

2.1.2   Uniqueness

In the last section we saw that such a fraction as $51/22$ can be expanded into a simple continued fraction, $51/22 = \langle 2, 3, 7 \rangle$. It can be verified that $51/22$ can also be expressed as $\langle 2, 3, 6, 1 \rangle$, but it turns out that these are the only two representations of $51/22$. In general, we note that the simple continued fraction expansion (2.1.3) has an alternate form,

$$\frac{u_0}{u_1} = \langle a_0, a_1, \ldots, a_{j-1}, a_j \rangle = \langle a_0, a_1, \ldots, a_{j-2}, a_{j-1}, a_j - 1, 1 \rangle. \qquad (2.1.4)$$

The following result [159] establishes that these are the only two simple continued fraction expansions of a fixed rational number.

Theorem 2.1.1. *If* $\langle a_0, a_1, \ldots, a_j \rangle = \langle b_0, b_1, \ldots, b_n \rangle$, *where these finite continued fractions are simple, and if* $a_j > 1$ *and* $b_n > 1$, *then* $j = n$ *and* $a_i = b_i$ *for* $i = 0, 1, \ldots, n$.

Proof. We write y_i for the continued fraction $\langle b_i, b_{i+1}, \ldots, b_n \rangle$ and observe that

$$y_i = \langle b_i, b_{i+1}, \ldots, b_n \rangle = b_i + \cfrac{1}{\langle b_{i+1}, b_{i+2}, \ldots, b_n \rangle} = b_i + \cfrac{1}{y_{i+1}}. \qquad (2.1.5)$$

Thus we have $y_i > b_i$ and $y_i > 1$ for $i = 1, 2, \ldots, n - 1$, and $y_n = b_n > 1$. Consequently, $b_i = [y_i]$ for all values of i in the range $0 \le i \le n$. The hypothesis that the continued fractions are equal can be written in the form $y_0 = \xi_0$, where we are using the notation of equation (2.1.3). Now the definition of ξ_i as u_i/u_{i+1} implies that $\xi_{i+1} > 1$ for all values of $i \geq 0$, and so $a_i = [\xi_i]$ for $0 \leq i \leq j$ by equation (2.1.2). It follows from $y_0 = \xi_0$ that, taking integral parts, $b_0 = [y_0] = [\xi_0] = a_0$. By equations (2.1.2) and (2.1.5) we get

$$\frac{1}{\xi_1} = \xi_0 - a_0 = y_0 - b_0 = \frac{1}{y_1}, \quad \xi_1 = y_1, \quad a_1 = [\xi_1] = [y_1] = b_1.$$

This gives us the start of a proof by induction. We now establish that $\xi_i = y_i$ and $a_i = b_i$ imply that $\xi_{i+1} = y_{i+1}$ and $a_{i+1} = b_{i+1}$. To see this, we again use equations (2.1.2) and (2.1.5) to write

$$\frac{1}{\xi_{i+1}} = \xi_i - a_i = y_i - b_i = \frac{1}{y_{i+1}},$$
$$\xi_{i+1} = y_{i+1}, \quad a_{i+1} = [\xi_{i+1}] = [y_{i+1}] = b_{i+1}.$$

It must also follow that the continued fractions have the same length, that is, that $j = n$. For suppose that, say, $j < n$. From the preceding argument we have $\xi_j = y_j$, $a_j = b_j$. But $\xi_j = a_j$ by (2.1.2) and $y_j > b_j$ by (2.1.5), and so we have a contradiction. If we had assumed $j > n$, a symmetrical contradiction would have arisen, and thus j must equal n, and the theorem is proved. □

Theorem 2.1.2. *Any finite simple continued fraction represents a rational number. Conversely, any rational number can be expressed as a finite simple continued fraction, and in exactly two ways.*

2.1.3 Infinite Continued Fractions

Let $a_0, a_1, a_2, \ldots$ be an infinite sequence of integers, all positive except perhaps a_0. We define two sequences of integers $\{h_n\}$ and $\{k_n\}$ inductively as follows:

$$h_{-2} = 0, \ h_{-1} = 1, \ h_i = a_i h_{i-1} + h_{i-2} \text{ for } i \geq 0$$
$$k_{-2} = 1, \ k_{-1} = 0, \ k_i = a_i k_{i-1} + k_{i-2} \text{ for } i \geq 0. \tag{2.1.6}$$

We note that $k_0 = 1$, $k_1 = a_1 k_0 \geq k_0$, $k_2 > k_1$, $k_3 > k_2$, etc., so that $1 = k_0 \leq k_1 < k_2 < k_3 < \cdots < k_n < \ldots$.

Theorem 2.1.3. *For any positive real number x,*

$$\langle a_0, a_1, \ldots, a_{n-1}, x \rangle = \frac{x h_{n-1} + h_{n-2}}{x k_{n-1} + k_{n-2}}.$$

Theorem 2.1.4. *If we define $r_n = \langle a_0, a_1, \ldots, a_n \rangle$ for all integers $n \geq 0$, then $r_n = h_n/k_n$.*

Theorem 2.1.5. *The equations*

$$h_i k_{i-1} - h_{i-1} k_i = (-1)^{i-1} \quad and \quad r_i - r_{i-1} = \frac{(-1)^{i-1}}{k_i k_{i-1}}$$

hold for $i \geq 1$. The identities

$$h_i k_{i-2} - h_{i-2} k_i = (-1)^i a_i \quad and \quad r_i - r_{i-2} = \frac{(-1)^i a_i}{k_i k_{i-2}}$$

hold for $i \geq 1$. The fraction h_i/k_i is reduced, that is $(h_i, k_i) = 1$.

Theorem 2.1.6. *The values r_n defined in Theorem 2.1.4 satisfy the infinite chain of inequalities $r_0 < r_2 < r_4 < r_6 < \cdots < r_7 < r_5 < r_3 < r_1$. Furthermore, $\lim_{n \to \infty} r_n$ exists, and for every $j \geq 0$, $r_{2j} < \lim_{n \to \infty} r_n < r_{2j+1}$.*

Proof. The identities of Theorem 2.1.5 for $r_i - r_{i-1}$ and $r_i - r_{i-2}$ imply that $r_{2j} < r_{2j+2}$, $r_{2j-1} > r_{2j+1}$, and $r_{2j} < r_{2j-1}$, because the k_i are positive for $i \geq 0$ and the a_i are positive for $i \geq 1$. Thus we have $r_0 < r_2 < r_4 < \ldots$ and $r_1 > r_3 > r_5 > \ldots$. To prove that $r_{2n} < r_{2j-1}$, we put the previous results together in the form

$$r_{2n} < r_{2n+2j} < r_{2n+2j-1} \leq r_{2j-1}.$$

The sequence $r_0, r_2, r_4, \ldots$ is monotonically increasing and is bounded above by r_1, and so has a limit. Analogously, the sequence $r_1, r_3, r_5, \ldots$ is monotonically decreasing and is bounded below by r_0, and so has a limit. These two limits are equal because, by Theorem 2.1.5, the difference $r_i - r_{i-1}$ tends to zero as i tends to infinity, since the integers k_i are increasing with i. Another way of looking at this to observe that (r_0, r_1), (r_2, r_3), $(r_4, r_5), \ldots$ is a chain of nested intervals defining a real number, namely $\lim_{n \to \infty} r_n$. □

These theorems suggest the following definition.

Definition 2.1.1. An infinite sequence $a_0, a_1, a_2, \ldots$ of integers, all positive except perhaps for a_0, determines an infinite simple continued fraction $\langle a_0, a_1, a_2, \ldots \rangle$. The value of $\langle a_0, a_1, a_2, \ldots \rangle$ is defined to be $\lim_{n \to \infty} \langle a_0, a_1, a_2, \ldots, a_n \rangle$.

This limit, being the same as $\lim_{n \to \infty} r_n$, exists by Theorem 2.1.6. Another way of writing this limit is $\lim_{n \to \infty} h_n/k_n$. The rational number $\langle a_0, a_1, \ldots, a_n \rangle = h_n/k_n = r_n$ is called the nth *convergent* to the infinite continued fraction. We say that the infinite continued fraction converges to the value $\lim_{n \to \infty} r_n$. In the case of a finite simple continued fraction $\langle a_0, a_1, \ldots, a_n \rangle$ we similarly call the number $\langle a_0, a_1, \ldots, a_m \rangle$ the mth *convergent* to $\langle a_0, a_1, \ldots, a_n \rangle$.

Theorem 2.1.7. *The value of any infinite simple continued fraction* $\langle a_0, a_1, a_2, \dots \rangle$
is irrational.

Proof. Writing θ for $\langle a_0, a_1, a_2, \dots \rangle$, we observe by Theorem 2.1.6 that θ lies
between r_n and r_{n+1}, so that $0 < |\theta - r_n| < |r_{n+1} - r_n|$. Multiplying by k_n, and
making use of the result from Theorem 2.1.5 that $|r_{n+1} - r_n| = (k_n k_{n+1})^{-1}$, we
have

$$0 < |k_n \theta - h_n| < \frac{1}{k_{n+1}}.$$

Now suppose that θ were rational, say $\theta = a/b$ with integers a and b, $b > 0$. Then
the above inequality would become, upon multiplication by b,

$$0 < |k_n a - h_n b| < \frac{b}{k_{n+1}}.$$

The integers k_n increase with n, so we could choose n sufficiently large so that
$b < k_{n+1}$. Then the integer $|k_n a - h_n b|$ would lie between 0 and 1, which is
impossible. $\qquad\square$

Lemma 2.1.8. *Let* $\theta = \langle a_0, a_1, a_2, \dots \rangle$ *be a simple continued fraction. Then*
$a_0 = [\theta]$. *Furthermore, if* θ_1 *denotes* $\langle a_1, a_2, a_3, \dots \rangle$, *then* $\theta = a_0 + 1/\theta_1$.

Proof. By Theorem 2.1.6 we see that $r_0 < \theta < r_1$, that is $a_0 < \theta < a_0 + 1/a_1$.
Now $a_1 \geq 1$, so we have $a_0 < \theta < a_0 + 1$, and hence $a_0 = [\theta]$. Also

$$\theta = \lim_{n \to \infty} \langle a_0, a_1, \dots, a_n \rangle = \lim_{n \to \infty} \left(a_0 + \frac{1}{\langle a_1, \dots, a_n \rangle} \right)$$

$$= a_0 + \frac{1}{\lim_{n \to \infty} \langle a_1, \dots, a_n \rangle} = a_0 + \frac{1}{\theta_1}.$$

$\qquad\square$

Theorem 2.1.9. *Two distinct infinite simple continued fractions converge to different values.*

Proof. Let us suppose that $\langle a_0, a_1, a_2, \dots \rangle = \langle b_0, b_1, b_2, \dots \rangle = \theta$. Then by
Lemma 2.1.8, $[\theta] = a_0 = b_0$ and

$$\theta = a_0 + \frac{1}{\langle a_1, a_2, \dots \rangle} = b_0 + \frac{1}{\langle b_1, b_2, \dots \rangle}.$$

Hence $\langle a_1, a_2, \dots \rangle = \langle b_1, b_2, \dots \rangle$. Repetition of the argument gives $a_1 = b_1$, and
so by induction $a_n = b_n$ for all n. $\qquad\square$

2.1.4   Irrational Numbers

We have shown that any infinite simple continued fraction represents an irrational number. Conversely, if we begin with an irrational number ξ, or ξ_0, we can expand it into an infinite simple continued fraction. To do this we define $a_0 = [\xi_0]$, $\xi_1 = 1/(\xi_0 - a_0)$, and next $a_1 = [\xi_1]$, $\xi_2 = 1/(\xi_1 - a_1)$, and so by an inductive definition

$$a_i = [\xi_i], \quad \xi_{i+1} = \frac{1}{\xi_i - a_i}. \tag{2.1.7}$$

The a_i are integers by definition, and the ξ_i are all irrational, since the irrationality of ξ_1 is implied by that of ξ_0, that of ξ_2 by that of ξ_1, and so on. Furthermore, $a_i \geq 1$ for $i \geq 1$ because $a_{i-1} = [\xi_{i-1}]$ and the fact that ξ_{i-1} is irrational implies that

$$a_{i-1} < \xi_{i-1} < 1 + a_{i-1}, \quad 0 < \xi_{i-1} - a_{i-1} < 1,$$

$$\xi_i = \frac{1}{\xi_{i-1} - a_{i-1}} > 1, \quad a_i = [\xi_i] \geq 1.$$

Next we use repeated application of (2.1.7) in the form $\xi_i = a_i + 1/\xi_{i+1}$ to get the chain

$$\xi = \xi_0 = a_0 + \frac{1}{\xi_1} = \langle a_0, \xi_1 \rangle$$

$$= \left\langle a_0, a_1 + \frac{1}{\xi_2} \right\rangle = \langle a_0, a_2, \xi_2 \rangle$$

$$= \left\langle a_0, a_1, \ldots, a_{m-2}, a_{m-1} + \frac{1}{\xi_m} \right\rangle$$

$$= \langle a_0, a_1, \ldots, a_{m-1}, \xi_m \rangle.$$

This suggests, but does not establish, that ξ is the value of the infinite continued fraction $\langle a_0, a_1, a_2, \ldots \rangle$ determined by the integers a_i.

To prove this we use Theorem 2.1.3 to write

$$\xi = \langle a_0, a_1, \ldots, a_{n-1}, \xi_n \rangle = \frac{\xi_n h_{n-1} + h_{n-2}}{\xi_n k_{n-1} + k_{n-2}} \tag{2.1.8}$$

with the h_i and k_i defined as in (2.1.6). By Theorem 2.1.5 we get

$$\xi - r_{n-1} = \xi - \frac{h_{n-1}}{k_{n-1}} = \frac{\xi_n h_{n-1} + h_{n-2}}{\xi_n k_{n-1} + k_{n-2}} - \frac{h_{n-1}}{k_{n-1}}$$

$$= \frac{-(h_{n-1}k_{n-2} - h_{n-2}k_{n-1})}{k_{n-1}(\xi_n k_{n-1} + k_{n-2})} = \frac{(-1)^{n-1}}{k_{n-1}(\xi_n k_{n-1} + k_{n-2})}. \tag{2.1.9}$$

This fraction tends to zero as n tends to infinity because the integers k_n are increasing with n, and ξ_n is positive. Hence $\xi - r_{n-1}$ tends to zero as n tends to infinity and then, by Definition 2.1.1,

$$\xi = \lim_{n\to\infty} r_n = \lim_{n\to\infty} \langle a_0, a_1, \ldots, a_n \rangle = \langle a_0, a_1, a_2, \ldots \rangle.$$

We summarize the results of the last two sections in the following theorem.

Theorem 2.1.10. *Any irrational number ξ is uniquely expressible, by the procedure that gave equations (2.1.7), as an infinite simple continued fraction $\langle a_0, a_1, a_2, \ldots \rangle$. Conversely, any such continued fraction determined by integers a_i that are positive for all $i > 0$ represents an irrational number, ξ. The finite simple continued fraction $\langle a_0, a_1, \ldots, a_n \rangle$ has the rational value $h_n/k_n = r_n$, and is called the nth convergent to ξ. Equations (2.1.6) relate the h_i and k_i to the a_i. For $n = 0, 2, 4, \ldots$ these convergents form a monotonically sequence with ξ as a limit. Similarly, for $n = 1, 3, 5, \ldots$ the convergents form a monotonically decreasing sequence tending to ξ. The denominators k_n of the convergents are an increasing sequence of positive integers for $n > 0$. Finally, with ξ_i defined by (2.1.7), we have $\langle a_0, a_1, \ldots \rangle = \langle a_0, a_1, \ldots, a_{n-1}, \xi_n \rangle$ and $\xi_n = \langle a_n, a_{n+1}, a_{n+2}, \ldots \rangle$.*

Proof. Only the last equation is new, and it becomes obvious if we apply to ξ_n the process described at the opening of this section. $\square$

Example 1. Let us expand $\sqrt{5}$ as an infinite simple continued fraction.
    We see that

$$\sqrt{5} = 2 + (\sqrt{5} - 2) = 2 + 1/(\sqrt{5} + 2)$$

and

$$\sqrt{5} + 2 = 4 + (\sqrt{5} - 2) = 4 + 1/(\sqrt{5} + 2).$$

In view of the repetition of $1/(\sqrt{5} + 2)$, we obtain $\sqrt{5} = \langle 2, 4, 4, 4, \ldots \rangle$.

2.1.5 Approximations to Irrational Numbers

Continuing to use the notation on the preceding sections, we now show that the convergents h_n/k_n form a sequence of "best" rational approximations to the irrational number ξ.

Theorem 2.1.11. *We have for any $n \geq 0$,*

$$\left| \xi - \frac{h_n}{k_n} \right| < \frac{1}{k_n k_{n+1}} \quad and \quad |\xi k_n - h_n| < \frac{1}{k_{n+1}}.$$

Proof. The second inequality follows from the first by multiplication by k_n. By (2.1.9) and (2.1.7) we have

$$\left|\xi - \frac{h_n}{k_n}\right| = \frac{1}{k_n(\xi_{n+1}k_n + k_{n-1})} < \frac{1}{k_n(a_{n+1}k_n + k_{n-1})}.$$

Using (2.1.6), we replace $a_{n+1}k_n + k_{n-1}$ by k_{n+1} to obtain the first inequality. □

Theorem 2.1.12. *The convergents h_n/k_n are successively closer to ξ, that is*

$$\left|\xi - \frac{h_n}{k_n}\right| < \left|\xi - \frac{h_{n-1}}{k_{n-1}}\right|.$$

In fact the stronger inequality $|\xi k_n - h_n| < |\xi k_{n-1} - h_{n-1}|$ holds.

Proof. We use $k_{n-1} \le k_n$ to write

$$\left|\xi - \frac{h_n}{k_n}\right| = \frac{1}{k_n}|\xi k_n - h_n| < \frac{1}{k_n}|\xi k_{n-1} - h_{n-1}|$$

$$\le \frac{1}{k_{n-1}}|\xi k_{n-1} - h_{n-1}| = \left|\xi - \frac{h_{n-1}}{k_{n-1}}\right|.$$

Now to prove the stronger inequality we observe that $a_n + 1 > \xi_n$ by (2.1.7), and so by (2.1.6), we have

$$\xi_n k_{n-1} + k_{n-2} < (a_n + 1)k_{n-1} + k_{n-2}$$

$$= k_n + k_{n-1} \le a_{n+1}k_n + k_{n-1} = k_{n+1}.$$

This inequality and (2.1.9) imply that

$$\left|\xi - \frac{h_{n-1}}{k_{n-1}}\right| = \frac{1}{k_{n-1}(\xi_n k_{n-1} + k_{n-2})} > \frac{1}{k_{n-1}k_{n+1}}.$$

We multiply by k_{n-1} and use Theorem 2.1.11 to get

$$|\xi k_{n-1} - h_{n-1}| > \frac{1}{k_{n+1}} > |\xi k_n - h_n|.$$

□

The convergent h_n/k_n is the best approximation to ξ of all the rational fractions with denominator k_n or less. The following theorem states this in a different way. For the proof we refer to [159].

Theorem 2.1.13. *If a/b is a rational number with positive denominator such that $|\xi - a/b| < |\xi - h_n/k_n|$ for some $n \geq 1$, then $b > k_n$. In fact if $|\xi b - a| < |\xi k_n - h_n|$ for some $n \geq 0$, then $b \geq k_{n+1}$.*

Theorem 2.1.14. *Let ξ denote any irrational number. If there is a rational number a/b with $b \geq 1$ such that*

$$\left| \xi - \frac{a}{b} \right| < \frac{1}{2b^2},$$

then a/b equals one of the convergents of the simple continued fraction expansion of ξ.

Theorem 2.1.15. *The nth convergent of $1/x$ is the reciprocal of the $(n-1)$st convergent of x if x is any real number greater than 1.*

2.1.6   Best Possible Approximations

Theorem 2.1.11 provides another method of proving the following well-known result (see [159, p. 302]). If ξ is real and irrational, there are infinitely many distinct rational numbers a/b such that

$$\left| \xi - \frac{a}{b} \right| < \frac{1}{b^2}.$$

Indeed, from Theorem 2.1.11 we can replace k_{n+1} by the smaller integer k_n to get the weaker, but still correct, inequality

$$\left| \xi - \frac{h_n}{k_n} \right| < \frac{1}{k_n^2}.$$

We can also use continued fractions to get different proofs of the following result of Hurwitz [159, pp. 304–305]:

Given an irrational number ξ, there exist infinitely many different rational numbers h/k such that

$$\left| \xi - \frac{h}{k} \right| < \frac{1}{\sqrt{5}k^2}$$

and the constant $\sqrt{5}$ is the best possible. The following auxiliary result is a simple consequence of the sign of the quadratic function.

Lemma 2.1.16. *If x is real, $x > 1$, and $x + x^{-1} < \sqrt{5}$, then $x < \frac{1}{2}(\sqrt{5}+1)$ and $x^{-1} > \frac{1}{2}(\sqrt{5}-1)$.*

Theorem 2.1.17 (Hurwitz). *Given any irrational number ξ, there exist infinitely many rational numbers h/k such that*

$$\left| \xi - \frac{h}{k} \right| < \frac{1}{\sqrt{5}k^2}. \tag{2.1.10}$$

Proof. The idea is to establish that, of every three consecutive convergents of the simple continued fraction expansion of ξ, at least one satisfies the inequality (2.1.10).

Let q_n denote k_n/k_{n-1}. We first prove that

$$q_j + q_j^{-1} < \sqrt{5} \tag{2.1.11}$$

if (2.1.10) is false for both $h/k = h_{j-1}/k_{j-1}$ and $h/k = h_j/k_j$. Suppose (2.1.10) is false for these two values of h/k. We have

$$\left| \xi - \frac{h_{j-1}}{k_{j-1}} \right| + \left| \xi - \frac{h_j}{k_j} \right| \geq \frac{1}{\sqrt{5}k_{j-1}^2} + \frac{1}{\sqrt{5}k_j^2}.$$

But ξ lies between h_{j-1}/k_{j-1} and h_j/k_j and hence we find, using Theorem 2.1.5, that

$$\left| \xi - \frac{h_{j-1}}{k_{j-1}} \right| + \left| \xi - \frac{h_j}{k_j} \right| = \left| \frac{h_{j-1}}{k_{j-1}} - \frac{h_j}{k_j} \right| = \frac{1}{k_{j-1}k_j}.$$

Combining these results we get

$$\frac{k_j}{k_{j-1}} + \frac{k_{j-1}}{k_j} \leq \sqrt{5}.$$

Since the left side is rational we actually have a strict inequality, and (2.1.11) follows.

Now suppose (2.1.10) is false for $h/k = h_i/k_i$, $i = n-1, n, n+1$. We then have (2.1.11) for both $j = n$ and $j = n+1$. By Lemma 2.1.16 we see that $q_n^{-1} > \frac{1}{2}(\sqrt{5}-1)$ and $q_{n+1} < \frac{1}{2}(\sqrt{5}+1)$, and, by (2.1.6) we find $q_{n+1} = a_{n+1} + q_n^{-1}$. This gives us

$$\frac{1}{2}(\sqrt{5}+1) > q_{n+1} = a_{n+1} + q_n^{-1} > a_{n+1} + \frac{1}{2}(\sqrt{5}-1)$$

$$\geq 1 + \frac{1}{2}(\sqrt{5}-1) = \frac{1}{2}(\sqrt{5}+1)$$

and this is a contradiction. □

Theorem 2.1.18. *The constant $\sqrt{5}$ in Theorem 2.1.17 is best possible, i.e., Theorem 2.1.17 does not hold if $\sqrt{5}$ is replaced by any larger value.*

Proof. It suffices to exhibit an irrational number ξ for which $\sqrt{5}$ is the largest possible constant. Consider the irrational ξ whose continued fraction expansion is $\langle 1, 1, 1, \ldots \rangle$. We see that

$$\xi = 1 + \frac{1}{\langle 1, 1, \ldots \rangle} = 1 + \frac{1}{\xi}, \quad \xi^2 = \xi + 1, \quad \xi = \frac{1}{2}(\sqrt{5} + 1).$$

Using (2.1.7) we can prove by induction that $\xi_i = (\sqrt{5} + 1)/2$ for all $i \geq 0$, for if $\xi_i = (\sqrt{5} + 1)/2$ then

$$\xi_{i+1} = (\xi_i - a_i)^{-1} = \left(\frac{1}{2}(\sqrt{5} + 1) - 1 \right)^{-1} = \frac{1}{2}(\sqrt{5} + 1).$$

A simple calculation yields $h_0 = k_0 = k_1 = 1$, $h_1 = k_2 = 2$. Equation (2.1.6) becomes $h_i = h_{i-1} + h_{i-2}$, $k_i = k_{i-1} + k_{i-2}$, and so by induction $k_n = h_{n-1}$ for $n \geq 1$. Hence we have

$$\lim_{n \to \infty} \frac{k_{n-1}}{k_n} = \lim_{n \to \infty} \frac{k_{n-1}}{h_{n-1}} = \frac{1}{\xi} = \frac{\sqrt{5} - 1}{2}$$

$$\lim_{n \to \infty} \left(\xi_{n+1} + \frac{k_{n-1}}{k_n} \right) = \frac{\sqrt{5} + 1}{2} + \frac{\sqrt{5} - 1}{2} = \sqrt{5}.$$

If c is any constant exceeding $\sqrt{5}$, then

$$\xi_{n+1} + \frac{k_{n-1}}{k_n} > c$$

holds for only a finite number of values of n. Thus, by (2.1.9),

$$\left| \xi - \frac{h_n}{k_n} \right| = \frac{1}{k_n^2(\xi_{n+1} + k_{n-1}/k_n)} < \frac{1}{ck_n^2}$$

holds for only a finite number of values of n. Thus there are only a finite number of rational numbers h/k satisfying $|\xi - h/k| < 1/(ck^2)$, because any such h/k is one of the convergents to ξ by Theorem 2.1.14. $\qquad \square$

2.1.7 Periodic Continued Fractions

An infinite simple continued fraction $\langle a_0, a_1, a_2, \ldots \rangle$ is said to be *periodic* if there is an integer n such that $a_r = a_{n+r}$ for all sufficiently large integers r. Thus a periodic continued fraction can be written in the form

$$\langle b_0, b_1, b_2, \ldots, b_j, a_0, a_1, \ldots, a_{n-1}, a_0, a_1, \ldots, a_{n-1}, \ldots \rangle$$
$$= \langle b_0, b_1, b_2, \ldots, b_j, \overline{a_0, a_1, \ldots, a_{n-1}} \rangle \qquad (2.1.12)$$

where the bar over the $a_0, a_1, \ldots, a_{n-1}$ indicates that this block of integers is repeated indefinitely. For example $\langle \overline{2,3} \rangle$ denotes $\langle 2, 3, 2, 3, 2, 3, \ldots \rangle$ and its value is easily computed. Writing θ for $\langle \overline{2,3} \rangle$ we have

$$\theta = 2 + \cfrac{1}{3 + \cfrac{1}{\theta}}.$$

This is a quadratic equation in θ, and we discard the negative root to obtain the value $\theta = (3 + \sqrt{15})/3$. As a second example consider $\langle 4, 1, \overline{2,3} \rangle$. Calling this ξ, we have $\xi = \langle 4, 1, \theta \rangle$, with θ as above, and so

$$\xi = 4 + (1 + \theta^{-1})^{-1} = 4 + \frac{\theta}{\theta + 1} = \frac{29 + \sqrt{15}}{7}.$$

These two examples illustrate the following result (see [159]).

Theorem 2.1.19. *Any periodic simple continued fraction is a quadratic irrational number, and conversely.*

Proof. Let us write ξ for the periodic continued fraction of (2.1.12) and θ for its purely periodic part,

$$\theta = \langle \overline{a_0, a_1, \ldots, a_{n-1}} \rangle = \langle a_0, a_1, \ldots, a_{n-1}, \theta \rangle.$$

Then equation (2.1.8) gives

$$\theta = \frac{\theta h_{n-1} + h_{n-2}}{\theta k_{n-1} + k_{n-2}}$$

and this is a quadratic equation in θ. Hence θ is either a quadratic irrational number or a rational number, but the latter is ruled out by Theorem 2.1.7. Now ξ can be written in terms on θ,

$$\xi = \langle b_0, b_1, \ldots, b_j, \theta \rangle = \frac{\theta m + m'}{\theta q + q'}$$

where m'/q' and m/q are the last two convergents to $\langle b_0, b_1, \ldots, b_j \rangle$. But θ is of the form $(a + \sqrt{b})/c$, and hence ξ is of similar form because, as with θ, we can rule out the possibility that ξ is rational.

To prove the converse, let us begin with any quadratic irrational ξ, or ξ_0, of the form $\xi = \xi_0 = (a + \sqrt{b})/c$, with integers $a, b, c > 0$, $c \neq 0$. The integer b is not a perfect square since ξ is irrational. We multiply numerator and denominator by $|c|$ to get

$$\xi_0 = \frac{ac + \sqrt{bc^2}}{c^2} \quad \text{or} \quad \xi_0 = \frac{-ac + \sqrt{bc^2}}{-c^2}$$

according as c is positive or negative. Thus we can write ξ in the form

$$\xi_0 = \frac{m_0 + \sqrt{d}}{q_0}$$

where $q_0 | (d - m_0^2), d, m_0$ and q_0 are integers, $q_0 \neq 0$, d not a perfect square. By writing ξ_0 in this form we can get a simple formulation of its continued fraction expansion $\langle a_0, a_1, a_2, \ldots \rangle$. We will prove that the equations

$$a_i = [\xi_i], \quad \xi_i = \frac{m_i + \sqrt{d}}{q_i}$$

$$m_{i+1} = a_i q_i - m_i, \quad q_{i+1} = \frac{d - m_{i+1}^2}{q_i} \tag{2.1.13}$$

define infinite sequences of integers m_i, q_i, a_i, and irrationals ξ_i in such a way that equations (2.1.7) hold, and hence we will have the continued fraction expansion of ξ_0.

In the first step, we start with ξ_0, m_0, q_0 as determined above, and we let $a_0 = [\xi_0]$. If ξ_i, m_i, q_i, a_i are known, then we take $m_{i+1} = a_i q_i - m_i$, $q_{i+1} = (d - m_{i+1}^2)/q_i$, $\xi_{i+1} = (m_{i+1} + \sqrt{d})/q_{i+1}$, $a_{i+1} = [\xi_{i+1}]$. That is, (2.1.13) actually does determine sequences ξ_i, m_i, q_i, a_i.

Now we use induction to prove that the m_i and q_i are integers such that $q_i \neq 0$ and $q_i | (d - m_i^2)$.

Next we can verify that

$$\xi_i - a_i = \frac{-a_i q_i + m_i + \sqrt{d}}{q_i} = \frac{\sqrt{d} - m_{i+1}}{q_i} = \frac{d - m_{i+1}^2}{q_i(\sqrt{d} + m_{i+1})}$$

$$= \frac{q_{i+1}}{\sqrt{d} + m_{i+1}} = \frac{1}{\xi_{i+1}}$$

which verifies (2.1.7) and so we have proved that $\xi_0 = \langle a_0, a_1, a_2, \ldots \rangle$, with a_i defined by (2.1.13).

Let $\xi_i' = (m_i - \sqrt{d})/q_i$, the conjugate of ξ_i. We get the equation

$$\xi_0' = \frac{\xi_n' h_{n-1} + h_{n-2}}{\xi_n' k_{n-1} + k_{n-2}}$$

by taking conjugates in (2.1.8). Solving for ξ_n' we obtain

$$\xi_n' = -\frac{k_{n-2}}{k_{n-1}} \left(\frac{\xi_0' - h_{n-2}/k_{n-2}}{\xi_0' - h_{n-1}/k_{n-1}} \right).$$

As n tends to infinity, both h_{n-1}/k_{n-1} and h_{n-2}/k_{n-2} tend to ξ_0, which is different from ξ_0', and hence the fraction in parentheses tends to 1. Thus for sufficiently large n, say $n > N$ where N is fixed, the fraction in parentheses is positive, and ξ_n' is negative. But ξ_n is positive for $n \geq 1$ and hence $\xi_n - \xi_n' > 0$ and $n > N$. Applying (2.1.13) we see that this gives $2\sqrt{d}/q_n > 0$ and hence $q_r > 0$ for $n > N$.

It also follows from (2.1.13) that

$$q_n q_{n+1} = d - m_{n+1}^2 \leq d, \quad q_n \leq q_n q_{n+1} \leq d$$
$$m_{n+1}^2 < m_{n+1}^2 + q_n q_{n+1} = d, \quad |m_{n+1}| < \sqrt{d}$$

for $n > N$. Since d is a fixed positive integer we conclude that q_n and m_{n+1} can assume only a fixed number of possible values for $n > N$. Hence the ordered pairs (m_n, q_n) can assume only a fixed number of possible pair values for $n > N$, and so there are distinct integers j and k such that $m_j = m_k$ and $q_j = q_k$. We can suppose we have chosen j and k so that $j < k$. By (2.1.13) this implies that $\xi_j = \xi_k$ and hence that

$$\xi_0 = \langle a_0, a_1, \ldots, a_{j-1}, \overline{a_j, a_{j+1}, \ldots, a_{k-1}} \rangle,$$

and we are done. $\square$

The following result describes the subclass of real quadratic irrationals that have purely periodic continued fraction expansions, that is, expressions of the form $\langle \overline{a_0, a_1, \ldots, a_n} \rangle$ (see [159]).

Theorem 2.1.20. *The continued fraction expansion of the real quadratic irrational number ξ is purely periodic if and only if $\xi > 1$ and $-1 < \xi' < 0$, where ξ' denotes the conjugate of ξ.*

Proof. First we assume that $\xi > 1$ and $-1 < \xi' < 0$. As usual, we write ξ_0 for ξ and take conjugates in (2.1.7) to obtain

$$\frac{1}{\xi_{i+1}'} = \xi_i' - a_i. \tag{2.1.14}$$

Now $a_i \geq 1$ for all i, even for $i = 0$, since $\xi_0 > 1$. Hence if $\xi_i' < 0$, then $1/\xi_{i+1}' < -1$, and we have $-1 < \xi_{i+1}' < 0$. Since $-1 < \xi_0' < 0$, we see, by mathematical induction, that $-1 < \xi_i' < 0$ holds for all $i \geq 0$. Then, since $\xi_i' = a_i + 1/\xi_{i+1}'$ by (2.1.14), we have

$$0 < -\frac{1}{\xi'_{i+1}} - a_i < 1, \quad a_i = \left[-\frac{1}{\xi'_{i+1}}\right].$$

Now ξ is a quadratic irrational, so $\xi_j = \xi_k$ for some integers j and k with $0 < j < k$. Then we have $\xi'_j = \xi'_k$ and

$$a_{j-1} = \left[-\frac{1}{\xi'_j}\right] = \left[-\frac{1}{\xi'_k}\right] = a_{k-1}$$

$$\xi_{j-1} = a_{j-1} + \frac{1}{\xi_j} = a_{k-1} + \frac{1}{\xi_k} = \xi_{k-1}.$$

Thus $\xi_j = \xi_k$ implies $\xi_{j-1} = \xi_{k-1}$. A j-fold iteration of this implication gives us $\xi_0 = \xi_{k-j}$, and we have

$$\xi = \xi_0 = \langle \overline{a_0, a_1, \ldots, a_{k-j-1}} \rangle.$$

To prove the converse, let us assume that ξ is purely periodic, say $\xi = \langle \overline{a_0, a_1, \ldots, a_{n-1}} \rangle$. where $a_0, a_1, \ldots, a_{n-1}$ are positive integers. Then $\xi > a_0 \geq 1$. Also, by (2.1.8) we have

$$\xi = \langle a_0, a_1, \ldots, a_{n-1}, \xi \rangle = \frac{\xi h_{n-1} + h_{n-2}}{\xi k_{n-1} + k_{n-2}}.$$

Thus ξ satisfies the equation

$$f(x) = x^2 k_{n-1} + x(k_{n-2} - h_{n-1}) - h_{n-2} = 0.$$

This quadratic equation has two roots, ξ and its conjugate ξ'. Since $\xi > 1$, we need to only prove that $f(x)$ has a root between -1 and 0 in order to establish that $-1 < \xi' < 0$. We will do this by showing that $f(-1)$ and $f(0)$ have opposite signs. First we observe that $f(0) = -h_{n-2} < 0$ by (2.1.6), since $a_i > 0$ for $i \geq 0$. Next we see that for $n \geq 1$

$$f(-1) = k_{n-1} - k_{n-2} + h_{n-1} - h_{n-2}$$
$$= (k_{n-2} + h_{n-2})(a_{n-1} - 1) + k_{n-3} + h_{n-3} \geq k_{n-3} + h_{n-3} > 0.$$

$\square$

We now turn to the continued fraction expansion of $\sqrt{d}$ for a positive integer d not a perfect square. We get at this by considering the closely related irrational number $\sqrt{d} + [\sqrt{d}]$. This number satisfies the conditions of Theorem 2.1.20, and so its continued fraction is purely periodic,

$$\sqrt{d} + [\sqrt{d}] = \langle \overline{a_0, a_1, \ldots, a_{r-1}} \rangle = \langle a_0, \overline{a_1, \ldots, a_{r-1}, a_0} \rangle. \tag{2.1.15}$$

We can suppose that we have chosen r to be the smallest integer for which $\sqrt{d} + [\sqrt{d}]$ has an expansion of the form (2.1.15). Now we note that $\xi_i = \langle a_i, a_{i+1}, \dots \rangle$ is purely periodic for all values of i, and that $\xi_0 = \xi_r = \xi_{2r} = \dots$. Furthermore, $\xi_1, \xi_2, \dots, \xi_{r-1}$ are all different from ξ_0, since otherwise there would be a shorter period. Thus $\xi_i = \xi_0$ if and only if i is of the form mr.

Now we can start with $\xi_0 = \sqrt{d} + [\sqrt{d}]$, $q_0 = 1$, $m_0 = [\sqrt{d}]$ in (2.1.13) because $1 | (d - [\sqrt{d}]^2)$. Then, for all $j \geq 0$, we have

$$\frac{m_{jr} + \sqrt{d}}{q_{jr}} = \xi_{jr} = \xi_0 = \frac{m_0 + \sqrt{d}}{q_0} = [\sqrt{d}] + \sqrt{d}$$

$$m_{jr} - q_{jr}[\sqrt{d}] = (q_{jr} - 1)\sqrt{d} \qquad (2.1.16)$$

and hence $q_{jr} = 1$, since the left side is rational and $\sqrt{d}$ is irrational. Moreover $q_i = 1$ for no other values of the subscript i. For $q_i = 1$ implies $\xi_i = m_i + \sqrt{d}$, but ξ_i has a purely periodic expansion so that, by Theorem 2.1.20 we have $-1 < m_i - \sqrt{d} < 0$, $\sqrt{d} - 1 < m_i < \sqrt{d}$, and hence $m_i = [\sqrt{d}]$. Thus $\xi_i = \xi_0$ and i is a multiple of r.

We also establish that $q_i = -1$ does not hold for any i. For $q_i = -1$ implies $\xi_i = -m_i - \sqrt{d}$ by (2.1.13), and by Theorem 2.1.20 we would have $-m_i - \sqrt{d} > 1$ and $-1 < -m_i + \sqrt{d} < 0$. But this implies $\sqrt{d} < m_i < -\sqrt{d} - 1$, which is impossible.

Noting that $a_0 = [\sqrt{d} + [\sqrt{d}]] = 2[\sqrt{d}]$, we can now turn to the case $\xi = \sqrt{d}$. Using (2.1.15) we have

$$\sqrt{d} = -[\sqrt{d}] + (\sqrt{d} + [\sqrt{d}])$$

$$= -[\sqrt{d}] + \langle 2[\sqrt{d}], \overline{a_1, a_2, \dots, a_{r-1}, a_0} \rangle$$

$$= \langle [\sqrt{d}], \overline{a_1, a_2, \dots, a_{r-1}, a_0} \rangle$$

with $a_0 = 2[\sqrt{d}]$.

When we apply (2.1.13) to $\sqrt{d} + [\sqrt{d}]$, $q_0 = 1$, $m_0 = [\sqrt{d}]$ we have $a_0 = 2[\sqrt{d}]$, $m_1 = [\sqrt{d}]$, $q_1 = d - [\sqrt{d}]^2$. But we can also apply (2.1.13) to $\sqrt{d}$ with $q_0 = 1$, $m_0 = 0$, and we find $a_0 = [\sqrt{d}]$, $m_1 = [\sqrt{d}]$, $q_1 = d - [\sqrt{d}]^2$. The value of a_0 is different, but the values of m_1, and of q_1, are the same in both cases. Since $\xi_i = (m_i + \sqrt{d})/q_i$ we see that further application of (2.1.13) yields the same values for the a_i, for the m_i, and for the q_i, in both cases. In other words, the expansions of $\sqrt{d} + [\sqrt{d}]$ and $\sqrt{d}$ differ only in the values of a_0 and m_0. Stating our results explicitly for the case $\sqrt{d}$, we have the following theorem.

Theorem 2.1.21. *If the positive integer d is not a perfect square, the simple continued fraction expansion of $\sqrt{d}$ has the form*

$$\sqrt{d} = \langle a_0, \overline{a_1, a_2, \dots, a_{r-1}, 2a_0} \rangle$$

with $a_0 = [\sqrt{d}]$. Furthermore, with $\xi_0 = \sqrt{d}$, $q_0 = 1$, $m_0 = 0$, in equations (2.1.13), we have $q_i = 1$ if and only if $r|i$, and $q_i = -1$ holds for no subscript i. Here r denotes the length of the shortest period in the expansion of $\sqrt{d}$.

2.2 Units and Norms in Quadratic Rings

2.2.1 Quadratic Rings

Let R be the commutative ring (see [42] and [54])

$$R = \{m + n\sqrt{D} : m, n \in \mathbb{Z}\} \tag{2.2.1}$$

where D is a positive that is not a perfect square, endowed with the standard operations induced from the ring of integers $(\mathbb{Z}, +, \cdot)$. An element $\varepsilon \in R$ is called a *unit* in R if it is inversable, that is there exists $\varepsilon_1 \in R$ such that $\varepsilon\varepsilon_1 = \varepsilon_1\varepsilon = 1$. Two elements $\alpha, \beta \in R$ are said to be *divisibility associated* if there exists a unit $\varepsilon \in R$ such that $\alpha = \varepsilon\beta$. We will adopt the notation $\alpha \sim \beta$ to indicate that α and β have the property above. It is not difficult to see that "$\sim$" is an equivalence relation.

If $\mu \in R$, $\mu = a + b\sqrt{D}$, we will denote by $\overline{\mu}$ the element $\overline{\mu} = a - b\sqrt{D}$ and will call it the *conjugate* of μ.

2.2.2 Norms in Quadratic Rings

Let us denote by $N : R \to \mathbb{Z}$ the following function: if $\mu = a + b\sqrt{D}$, then

$$N(\mu) = a^2 - Db^2 = \mu \cdot \overline{\mu}. \tag{2.2.2}$$

Proposition 2.2.1 (N Is Multiplicative). *For all $\mu_1, \mu_2 \in R$, the following relation holds:*

$$N(\mu_1\mu_2) = N(\mu_1)N(\mu_2).$$

Proof. If $\mu_1 = m_1 + n_1\sqrt{D}$ and $\mu_2 = m_2 + n_2\sqrt{D}$, then we have

$$\mu_1\mu_2 = (m_1m_2 + Dn_1n_2) + (m_1n_2 + m_2n_1)\sqrt{D}$$

and

$$N(\mu_1\mu_2) = (m_1m_2 + Dn_1n_2)^2 - D(m_1n_2 + m_2n_1)^2$$
$$= m_1^2m_2^2 + D^2n_1^2n_2^2 - Dm_1^2n_2^2 - Dm_2^2n_1^2 = m_1^2(m_2^2 - Dn_2^2) - Dn_1^2(m_2^2 - dn_2^2)$$
$$= (m_1^2 - Dn_1^2)(m_2^2 - Dn_2^2) = N(\mu_1)N(\mu_2).$$

$\square$

Proposition 2.2.2. *An element $\varepsilon \in R$ is an unit in R if and only if $N(\varepsilon) = \pm 1$.*

Proof. If ε is a unit in R, then there exists $\varepsilon_1 \in R$ such that $\varepsilon\varepsilon_1 = 1$. Then from Proposition 2.2.1, $N(\varepsilon)N(\varepsilon_1) = N(1) = 1^2 - D0^2 = 1$. Since $N(\varepsilon)$ and $N(\varepsilon_1)$ are integers, it follows that $N(\varepsilon) = \pm 1$. Conversely, if $N(\varepsilon) = \pm 1$, then (2.2.2) yields $\varepsilon\overline{\varepsilon} = \pm 1$. If $N(\varepsilon) = 1$, then $\varepsilon\overline{\varepsilon} = 1$ and if $N(\varepsilon) = -1$, then $\varepsilon(-\overline{\varepsilon}) = 1$. Both cases show that ε is a unit in R. $\square$

Theorem 2.2.3. *For any integer a, the cardinal number of the set*

$$S = \{\alpha \in R : N(\alpha) = a \text{ and } \alpha \nsim \beta \text{ for all } \beta \in R, \ \beta \neq \alpha\} \qquad (2.2.3)$$

is finite and does not exceed a^2.

Proof. If $a = 0$, then the cardinal number of S is 1. We may assume now that a is nonzero. Let $\alpha, \beta \in S$ such that $\alpha \neq \beta$ and $\alpha \equiv \beta \pmod{a}$. This means that there exists $\gamma \in R$ such that $\alpha - \beta = a\gamma$.

From the definition of the set S it follows that $a = N(\alpha) = N(\beta)$, hence $\alpha - \beta = a\gamma = N(\alpha)\gamma = N(\beta)\gamma$.

Now embed the ring R into the field $\mathbb{Q}(\sqrt{D}) = \{r + s\sqrt{D} : r, s \in \mathbb{Q}\}$. Since $N(\alpha) = N(\beta) = a \neq 0$, we have $\alpha, \beta \neq 0$ and

$$\frac{\alpha}{\beta} = \frac{\beta + a\gamma}{\beta} = 1 + \frac{N(\beta)\gamma}{\beta} = 1 + \frac{\beta\overline{\beta}\gamma}{\beta} = 1 + \overline{\beta}\gamma$$

and

$$\frac{\beta}{\alpha} = \frac{\alpha - a\gamma}{\alpha} = 1 - \frac{N(\alpha)\gamma}{\alpha} = 1 - \frac{\alpha\overline{\alpha}\gamma}{\alpha} = 1 - \overline{\alpha}\gamma.$$

The computations above show that

$$\frac{\alpha}{\beta} - \frac{\beta}{\alpha} = (\overline{\beta} - \overline{\alpha})\gamma$$

hence $\dfrac{\alpha}{\beta}, \dfrac{\beta}{\alpha} \in R$ and $\alpha \sim \beta$, in contradiction with the definition of S. It follows that $\alpha \equiv \beta \pmod{a}$, for all $\alpha, \beta \in S$.

On the other hand, it is not difficult to see that for all b in R there exist positive integers m, n such that $0 \leq m < |a|, 0 \leq n < |a|$, and $b \equiv m + n\sqrt{D} \pmod{a}$.

The considerations above show that the mapping

$$S \rightarrow \{0, 1, 2, \ldots, |a| - 1\} \times \{0, 1, 2, \ldots, |a| - 1\}$$

given by $\alpha \rightarrow (m, n)$, where $0 \leq m, n \leq |a| - 1$, $\alpha \equiv m + n\sqrt{D} \pmod{a}$, is one-to-one.

This means that the set S is finite and its cardinal number is less or equal to a^2. $\square$

Proposition 2.2.4 (The Conjugate Is Multiplicative). *For all $\mu_1, \mu_2 \in R$, the following relation holds:*

$$\overline{\mu_1 \mu_2} = \overline{\mu}_1 \overline{\mu}_2. \tag{2.2.4}$$

Proof. If $\mu_1 = m_1 + n_1 \sqrt{D}$ and $\mu_2 = m_2 + n_2 \sqrt{D}$, then

$$\mu_1 \mu_2 = (m_1 m_2 + D n_1 n_2) + (m_1 n_2 + m_2 n_1) \sqrt{D}$$

and

$$\overline{\mu_1 \mu_2} = (m_1 m_2 + D n_1 n_2) - (m_1 n_2 + m_2 n_1) \sqrt{D}$$
$$= (m_1 - n_1 \sqrt{D})(m_2 - n_2 \sqrt{D}) = \overline{\mu}_1 \overline{\mu}_2.$$

$\square$

Remark. Proposition 2.2.4 gives another proof of the fact that N is multiplicative. Indeed, we have

$$N(\mu_1 \mu_2) = (\mu_1 \mu_2)(\overline{\mu_1 \mu_2}) = (\mu_1 \mu_2)(\overline{\mu}_1 \overline{\mu}_2) = (\mu_1 \overline{\mu}_1)(\mu_2 \overline{\mu}_2) = N(\mu_1)N(\mu_2).$$

Chapter 3
Pell's Equation

3.1 History and Motivation

Euler, after a cursory reading of Wallis's *Opera Mathematica*, mistakenly attributed the first serious study of nontrivial solutions to equations of the form $x^2 - Dy^2 = 1$, where $x \neq 1$ and $y \neq 0$, to John Pell. However, there is no evidence that Pell, who taught at the University of Amsterdam, had ever considered solving such equations. They should be probably called Fermat's equations, since it was Fermat who first investigated properties of nontrivial solutions of such equations. Nevertheless, Pellian equations have a long and rich history and can be traced back to the Greeks. For many details we refer to the books [75] and [212] (see also the reference [22, pp. 118–120]). Theon of Smyrna used x/y to approximate $\sqrt{2}$, where x and y were integral solutions to $x^2 - 2y^2 = 1$. In general, if $x^2 = Dy^2 + 1$, then $x^2/y^2 = D + 1/y^2$. Hence, for y large, x/y is a good approximation of $\sqrt{D}$, a fact well known to Archimedes.

The famous Archimedes's *problema bovinum* can be reduced to a such equation and it took two thousand years to solve (see [65]).

More precisely, it is reduced to the Pell's equation $x^2 - 4729494y^2 = 1$. The least positive solution, for which y has 41 digits, was discovered by Carl Amthov in 1880. For a nice presentation of the story of this problem we refer to the book [212].

In *Arithmetica*, Diophantus asks for rational solutions to equations of the type $x^2 - Dy^2 = 1$. In the case where $D = m^2 + 1$, Diophantus offered the integral solution $x = 2m^2 + 1$ and $y = 2m$. Pell type equations are also found in Hindu mathematics. In the fourth century, the Indian mathematician Baudhayana noted that $x = 577$ and $y = 408$ is a solution of $x^2 - 2y^2 = 1$ and used the fraction $\frac{577}{408}$ to approximate $\sqrt{2}$. In the seventh century, Brahmagupta considered solutions to the Pell's equation $x^2 - 92y^2 = 1$, the smallest solution being $x = 1151$ and $y = 120$. In the twelfth century, the Hindu mathematician Bhaskara found the least

© Springer Science+Business Media New York 2015
T. Andreescu, D. Andrica, *Quadratic Diophantine Equations*,
Developments in Mathematics 40, DOI 10.1007/978-0-387-54109-9_3

positive solution to the Pell's equation $x^2 - 61y^2 = 1$ to be $x = 226153980$ and $y = 1766319049$.

In 1657, Fermat stated without proof that if D is positive and not a perfect square, then Pell's equation has an infinite number of solutions. For if (x, y) is a solution to $x^2 - Dy^2 = 1$, then we have $1^2 = (x^2 - Dy^2)^2 = (x^2 + Dy)^2 - (2xy)^2 D$. Thus, $(x^2 + Dy, 2xy)$ is also a solution to $x^2 - Dy^2 = 1$. Therefore, if Pell's equation has a solution, then it has infinitely many.

In 1657, Fermat challenged William Brouncker and John Wallis to find integral solutions to the equations $x^2 - 151y^2 = 1$ and $x^2 - 313y^2 = -1$. He cautioned them not to submit rational solutions for even the lowest type of arithmetician could devise such answers. Wallis replied with $(1728148040, 140634693)$ as a solution to the first equation.

In 1770 Euler was looking for positive integers m and n such that $n(n + 1)/2 = m^2$. To accomplish this, he multiplied both sides of the latter equation by 8 and added 1 to obtain $(2n + 1)^2 = 8m^2 + 1$. He let $x = 2n + 1$ and $y = 2m$ so that $x^2 - 2y^2 = 1$. Solutions to this Pell's equation produce square-triangular numbers since we have

$$\frac{\left(\dfrac{x-1}{2}\right)\left(\dfrac{x-1}{2}+1\right)}{2} = \left(\frac{y}{2}\right)^2.$$

That is, the $\left(\dfrac{x-1}{2}\right)^{th}$ triangular number equals the $\left(\dfrac{y}{2}\right)^{th}$ square number. For example, from the solution $x = 3$ and $y = 2$, it follows that $m = n = 1$, yielding the square-triangular number 1. A natural question arises. Does the method generate all square-triangular numbers? If one is more methodical about how one obtains the solutions, one can see that it does.

Since $1 = x^2 - 2y^2 = (x - y\sqrt{2})(x + y\sqrt{2})$, it follows that

$$\begin{aligned} 1 = 1^2 &= (x - y\sqrt{2})^2(x + y\sqrt{2})^2 \\ &= ((2y^2 + x^2) - 2xy\sqrt{2})((2y^2 + x^2) + 2xy\sqrt{2}) \\ &= (2y^2 + x^2)^2 - 2(2xy)^2. \end{aligned}$$

Thus, if (x, y) is a solution to $1 = x^2 - 2y^2$, then so is $(2y^2 + x^2, 2xy)$. For example, the solution $(3, 2)$ generates the solution

$$(2 \cdot 2^2 + 3^2, 2 \cdot 2 \cdot 3) = (17, 12).$$

The solution $(17, 12)$ generates the solution

$$(2 \cdot 12^2 + 17^2, 2 \cdot 12 \cdot 17) = (577, 408).$$

The square-triangular number generated by the solution $(2y^2 + x^2, 2xy)$ to $1 = x^2 - 2y^2$ is distinct from the square-triangular number generated by the solution (x, y). Therefore, there exist an infinite number of square-triangular numbers. Lagrange, in a series of papers presented to the Berlin Academy between 1768 and 1770, showed that a similar procedure will determine all solutions to $x^2 = Dy^2 + 1$, where D is positive and nonsquare. In 1766, Lagrange proved that the equation $x^2 = Dy^2 + 1$ has an infinite number of solutions whenever D is positive and not square.

The Diophantine quadratic equation

$$ax^2 + bxy + cy^2 + dx + ey + f = 0 \tag{3.1.1}$$

with integral coefficients a, b, c, d, e, f reduces in its main case to a Pell-type equation. Next, we will sketch the general method of reduction. The equation (3.1.1) represents a conic in the xOy Cartesian plane, therefore solving (3.1.1) in integers means finding all lattice points situated on this conic. We will solve the equation (3.1.1) by reducing the general equation of the conic to its canonical form. Following the ideas from [13, 14, 160, 168] we introduce the discriminant of the equation (3.1.1) by $\Delta = b^2 - 4ac$. When $\Delta < 0$, the conic defined by (3.1.1) is an ellipse and in this case the given equation has only a finite number of solutions. If $\Delta = 0$, then the conic given by (3.1.1) is a parabola. If $2ae - bd = 0$, then the equation (3.1.1) becomes $(2ax + by + d)^2 = d^2 - 4af$ and it is not difficult to solve. In the case $2ae - bd \neq 0$, by performing the substitutions $X = 2ax + by + d$ and $Y = (4ae - 2bd)y + 4af - d^2$, the equation (3.1.1) reduces to $X^2 + Y = 0$ which is also easy to solve. The most interesting case is $\Delta > 0$, when the conic defined by (3.1.1) is a hyperbola. Using a sequence of substitutions, the equation (3.1.1) reduces to a general Pell-type equation

$$X^2 - DY^2 = N. \tag{3.1.2}$$

To illustrate the process described above, we will consider the equation $2x^2 - 6xy + 3y^2 = -1$ (Berkely Math. Circle 2000–2001 Monthly Contest #4, Problem 4, [22, p. 120]). We notice that $\Delta = 12 > 0$, hence the corresponding conic is a hyperbola. The equation can be written as $x^2 - 3(y - x)^2 = 1$ and by performing the substitutions $X = x$ and $Y = y - x$, we reduce it to the Pell's equation $X^2 - 3Y^2 = 1$.

Finally, let us mention that other authors reduce the equation (3.1.1) to the form $Ax^2 + Bxy + Cy^2 = k$ (see, for example [203]). Formulas yielding an infinite set of integral solutions of the Diophantine equation $x^2 + bx + c = ky^2$ are given in [80].

3.2 The General Solution by Elementary Methods

We will present an elementary approach to solving the Pell's equation due to Lagrange (see, for example, [93, 112, 125, 126, 191, 198] and [212]). We will follow the presentation of our papers [13–15].

Theorem 3.2.1. *If D is a positive integer that is not a perfect square, then the equation*

$$u^2 - Dv^2 = 1 \qquad (3.2.1)$$

has infinitely many solutions in positive integers and the general solution is given by $(u_n, v_n)_{n \geq 0}$,

$$u_{n+1} = u_1 u_n + D v_1 v_n, \quad v_{n+1} = v_1 u_n + u_1 v_n, \qquad (3.2.2)$$

where (u_1, v_1) *is its fundamental solution, i.e., the minimal solution different from the trivial solution* $(u_0, v_0) = (1, 0)$.

Proof. First, we will prove that the equation (3.2.1) has a fundamental solution.

Let c_1 be an integer greater than 1. We will show that there exist integers $t_1, w_1 \geq 1$ such that

$$|t_1 - w_1\sqrt{D}| < \frac{1}{c_1}, \quad w_1 \leq c_1.$$

Indeed, considering $l_k = [k\sqrt{D} + 1]$, $k = 0, \ldots, c_1$, yields $0 < l_k - k\sqrt{D} \leq 1$, $k = 0, \ldots, c_1$, and since $\sqrt{D}$ is an irrational number, it follows that $l_{k'} \neq l_{k''}$ whenever $k' \neq k''$.

There exist $i, j, p \in \{0, 1, 2, \ldots, c_1\}$, $i \neq j$, $p \neq 0$, such that

$$\frac{p-1}{c_1} < l_i - i\sqrt{D} \leq \frac{p}{c_1} \quad \text{and} \quad \frac{p-1}{c_1} < l_j - j\sqrt{D} \leq \frac{p}{c_1}$$

because there are c_1 intervals of the form $\left(\frac{p-1}{c_1}, \frac{p}{c_1} \right)$, $p = 1, \ldots, c_1$ and $c_1 + 1$ numbers of the form $l_k - k\sqrt{D}$, $k = 0, \ldots, c_1$.

From the inequalities above it follows that $|(l_i - l_j) - (j - i)\sqrt{D}| < \frac{1}{c_1}$ and setting $|l_i - l_j| = t_1$ and $|j - i| = w_1$ yields $|t_1 - w_1\sqrt{D}| < \frac{1}{c_1}$ and $w_1 \leq c_1$.

Multiplying this inequality by $t_1 + w_1\sqrt{D} < 2w_1\sqrt{D} + 1$ gives

$$|t_1^2 - Dw_1^2| < 2\frac{w_1}{c_1}\sqrt{D} + \frac{1}{c_1} < 2\sqrt{D} + 1.$$

Choosing a positive integer $c_2 > c_1$ such that $|t_1 - w_1\sqrt{D}| > \frac{1}{c_2}$, we obtain positive integers t_2, w_2 with the properties

$$|t_2^2 - Dw_2^2| < 2\sqrt{D} + 1 \quad \text{and} \quad |t_1 - t_2| + |w_1 - w_2| \neq 0.$$

By continuing this procedure, we find a sequence of distinct pairs $(t_n, w_n)_{n \geq 1}$ satisfying the inequalities $|t_n^2 - Dw_n^2| < 2\sqrt{D} + 1$ for all positive integers n. It follows that the interval $(-2\sqrt{D} - 1, 2\sqrt{D} + 1)$ contains a nonzero integer k such that there exists a subsequence of $(t_n, w_n)_{n \geq 1}$ satisfying the equation $t^2 - Dw^2 = k$. This subsequence contains at least two pairs (t_s, w_s), (t_r, w_r) for which $t_s \equiv t_r (\mathrm{mod}\ |k|)$, $w_s \equiv w_r (\mathrm{mod}\ |k|)$, and $t_s w_r - t_r w_s \neq 0$, otherwise $t_s = t_r$ and $w_s = w_r$, in contradiction with $|t_s - t_r| + |w_s - w_r| \neq 0$ see [21] and [23] for general properties of congruences).

Let $t_0 = t_s t_r - Dw_s w_r$ and let $w_0 = t_s w_r - t_r w_s$. Then

$$t_0^2 - Dw_0^2 = k^2. \tag{3.2.3}$$

On the other hand, $t_0 = t_s t_r - Dw_s w_r \equiv t_s^2 - Dw_0^2 \equiv 0 (\mathrm{mod}\ |k|)$, and it follows immediately that $w_0 \equiv 0 (\mathrm{mod}\ |k|)$. The pair (u, v), where $u = \dfrac{t_0}{|k|}$, $v = \dfrac{w_0}{|k|}$ is a nontrivial solution to Pell's equation (3.2.1).

Let (u_1, v_1) be the least such solution, i.e., with u (and implicitly v) minimal.

We show now that the pair (u_n, v_n) defined by (3.2.2) satisfies Pell's equation (3.2.1). We proceed by induction with respect to n. Clearly, (u_1, v_1) is a solution to the equation (3.2.1). If (u_n, v_n) is a solution to this equation, then

$$u_{n+1}^2 - Dv_{n+1}^2 = (u_1 u_n + Dv_1 v_n)^2 - D(v_1 u_n + u_1 v_n)^2$$
$$= (u_1^2 - Dv_1^2)(u_n^2 - Dv_n^2) = 1,$$

i.e., the pair (u_{n+1}, v_{n+1}) is also a solution to the equation (3.2.1).

It is not difficult to see that for all positive integer n,

$$u_n + v_n\sqrt{D} = (u_1 + v_1\sqrt{D})^n. \tag{3.2.4}$$

Clearly, (3.2.4) also yields the trivial solution $(u_0, v_0) = (1, 0)$.

Let $z_n = u_n + v_n\sqrt{D} = (u_1 + v_1\sqrt{D})^n$ and note that $z_0 < z_1 < \cdots < z_n < \cdots$. We will prove now that all solutions to the equation (3.2.1) are of the form (3.2.4). Indeed, if the equation (3.2.1) had a solution (u, v) such that $z = u + v\sqrt{D}$ is not of the form (3.2.4), then $z_m < z < z_{m+1}$ for some integer m. Then $1 < (u + v\sqrt{D})(u_m - v_m\sqrt{D}) < u_1 + v_1\sqrt{D}$, and therefore $1 < (uu_m - Dvv_m) + (u_m v - uv_m)\sqrt{D} < u_1 + v_1\sqrt{D}$. On the other hand, $(uu_m - Dvv_m)^2 - D(u_m v - uv_m)^2 = (u^2 - Dv^2)(u_m^2 - Dv_m^2) = 1$, i.e., $(uu_m - Dvv_m, u_m v - uv_m)$ is a solution of (3.2.1) smaller than (u_1, v_1), in contradiction with the assumption that (u_1, v_1) is the minimal nontrivial solution. $\qquad\square$

Remarks. 1) The relations (3.2.1) could be written in the following useful matrix form

$$\begin{pmatrix} u_{n+1} \\ v_{n+1} \end{pmatrix} = \begin{pmatrix} u_1 & Dv_1 \\ v_1 & u_1 \end{pmatrix} \begin{pmatrix} u_n \\ v_n \end{pmatrix}$$

from where

$$\begin{pmatrix} u_n \\ v_n \end{pmatrix} = \begin{pmatrix} u_1 & Dv_1 \\ v_1 & u_1 \end{pmatrix}^n \begin{pmatrix} u_0 \\ v_0 \end{pmatrix}. \tag{3.2.5}$$

If

$$\begin{pmatrix} u_1 & Dv_1 \\ v_1 & u_1 \end{pmatrix}^n = \begin{pmatrix} a_n & b_n \\ c_n & d_n \end{pmatrix}$$

then it is well-known that each of a_n, b_n, c_n, d_n is a linear combination of λ_1^n, λ_2^n, where λ_1, λ_2 are the eigenvalues of the matrix $\begin{pmatrix} u_1 & Dv_1 \\ v_1 & u_1 \end{pmatrix}$. By using (3.2.5), after an easy computation it follows that

$$u_n = \frac{1}{2}[(u_1 + v_1\sqrt{D})^n + (u_1 - v_1\sqrt{D})^n],$$

$$v_n = \frac{1}{2\sqrt{D}}[(u_1 + v_1\sqrt{D})^n - (u_1 - v_1\sqrt{D})^n] \tag{3.2.6}$$

2) The solutions of Pell's equation given in one of the forms (3.2.4) or (3.2.6) may be used in the approximation of the square roots of positive integers that are not perfect squares. Indeed, if (u_n, v_n) are the solutions of the equation (3.2.1), then

$$u_n - v_n\sqrt{D} = \frac{1}{u_n + v_n\sqrt{D}}$$

and so

$$\frac{u_n}{v_n} - \sqrt{D} = \frac{1}{v_n(u_n + v_n\sqrt{D})} < \frac{1}{\sqrt{D}v_n^2} < \frac{1}{v_n^2}.$$

It follows that

$$\lim_{n\to\infty} \frac{u_n}{v_n} = \sqrt{D} \tag{3.2.7}$$

i.e., the fractions $\frac{u_n}{v_n}$ approximate $\sqrt{D}$ with an error less than $\frac{1}{v_n^2}$.

3) Consider the plane transformation $T : \mathbb{R}^2 \to \mathbb{R}^2$, given by

$$T(x, y) = (u_1 x + Dv_1 y, v_1 x + u_1 y),$$

where (u_1, v_1) is the fundamental solution of Pell's equation (3.2.1). Let $(T^n)_{n\geq 0}$ be the discrete dynamical system generated by transformation T, where T^n

denotes the nth iteration of T. The result in Theorem 3.2.1 shows that the orbit of point (u_0, v_0) of this dynamical system consists of lattice points on the hyperbola $x^2 - Dy^2 = 1$.

4) It is not difficult to find rational solutions to equation (3.2.1). Simply divide the relation

$$(r^2 + D)^2 - D(2r)^2 = (r^2 - D)^2$$

by $(r^2 - D)^2$ to obtain

$$u = \frac{r^2 + D}{r^2 - D}, \quad v = \frac{2r}{r^2 - D}, \quad r \in \mathbb{Q}.$$

In the next sections we will see how we can describe all rational solutions to (3.2.1).

5) Dirichlet in 1837 published explicit formulae giving some solutions of Pell's equations in terms of trigonometric functions. For example, for $D = 13$ he has obtained $x_1 + y_1\sqrt{13} = \eta^2$, where

$$\eta = \frac{\sin \dfrac{2\pi}{13} \sin \dfrac{5\pi}{13} \sin \dfrac{6\pi}{13}}{\sin \dfrac{\pi}{13} \sin \dfrac{3\pi}{13} \sin \dfrac{4\pi}{13}} \in \mathbb{Q}(\sqrt{13}).$$

6) Concerning Pell's equation there is the following conjecture [150]: *Let p be a prime $\equiv 3 \pmod 4$. Consider Pell's equation $u^2 - pv^2 = 1$ and its fundamental solution (u_1, v_1). Then $v_1 \not\equiv 0 \pmod p$.*

This has been verified for all such primes $p < 18000$. It has been shown that $v_1 \not\equiv 0 \pmod p$ if and only if $E_{\frac{p-3}{4}} \not\equiv 0 \pmod p$, where the Euler numbers E_n are defined by the powers series

$$\sec t = \sum_{n=0}^{\infty} \frac{E_n}{(2n)!} t^{2n}.$$

There is a similar conjecture when $p \equiv 1 \pmod 4$.

3.3 The General Solution by Continued Fractions

The approach in this section is based on the material contained in Chapter 2, Section 2.1. More specifically, the method we are going to present is based on expanding $\sqrt{D}$ into a continued fraction as in Theorem 2.1.21, with convergents h_n/k_n, and with q_n defined by equations (2.1.13) with $\xi_0 = \sqrt{D}$, $q_0 = 1$, $m_0 = 0$. Our presentation is based on [1, 46, 159] and [164].

Theorem 3.3.1. *If D is a positive integer not a perfect square, then $h_n^2 - Dk_n^2 = (-1)^{n-1}q_{n+1}$ for all integers $n \geq -1$.*

Proof. From equations (2.1.8) and (2.1.13) we have

$$\sqrt{D} = \xi_0 = \frac{\xi_{n+1}h_n + h_{n-1}}{\xi_{n+1}k_n + k_{n-1}} = \frac{(m_{n+1} + \sqrt{D})h_n + q_{n+1}h_{n-1}}{(m_{n+1} + \sqrt{D})k_n + q_{n+1}k_{n-1}}.$$

We simplify this equation and separate it into a rational and a purely irrational part much as we did in (2.1.16). Each part must be zero so we get two equations, and we can eliminate m_{n+1} from them. The final result is

$$h_n^2 - Dk_n^2 = (h_n k_{n-1} - h_{n-1}k_n)q_{n+1} = (-1)^{n-1}q_{n+1}$$

where we used Theorem 2.1.5 in the last step. $\square$

Corollary 3.3.2. *Taking r as the length of the period of the expansion of $\sqrt{D}$, as in Theorem 2.1.21, we have for $n \geq 0$,*

$$h_{nr-1}^2 - Dk_{nr-1}^2 = (-1)^{nr}q_{nr} = (-1)^{nr}.$$

With n even, this gives infinitely many solutions of $x^2 - Dy^2 = 1$ in integers, provided D is positive and not a perfect square.

It can be seen that Theorem 3.3.1 gives us solutions to (3.1.2) for certain values of N. In particular, Corollary 3.3.2 gives infinitely many solutions of $x^2 - Dy^2 = 1$ by the use of even values nr. Of course, if r is even, all values of nr are even. If r is odd, Corollary 3.3.2 gives infinitely many solutions to $x^2 - Dy^2 = -1$ by the use of odd integers $n \geq 1$. The next result shows that every solution to $x^2 - Dy^2 = \pm 1$ can be obtained from the continued fraction expansion of $\sqrt{D}$. But first we make this simple observation: Apart from such trivial solutions as $x = \pm 1$, $y = 0$ of $x^2 - Dy^2 = 1$, all solutions to $x^2 - Dy^2 = N$ fall into sets of four by all combinations of signs $\pm x$, $\pm y$. Hence it is sufficient to discuss the positive solutions $x > 0$, $y > 0$.

Theorem 3.3.3. *Let D be a positive integer not a perfect square, and let the convergents to the continued expansion of $\sqrt{D}$ be h_n/k_n. Let the integer N satisfy $|N| < \sqrt{D}$. Then any positive solution $x = s$, $y = t$ to $x^2 - Dy^2 = N$, with $\gcd(s,t) = 1$, satisfies $s = h_n$, $t = k_n$ for some positive integer n.*

Proof. Let E and M be positive integers such that $\gcd(E,M) = 1$ and $E^2 - \rho M^2 = \sigma$, where $\sqrt{\rho}$ is irrational and $0 < \sigma < \sqrt{\rho}$. Here ρ and σ are real numbers, not necessarily integers. Then

$$\frac{E}{M} - \sqrt{\rho} = \frac{\sigma}{M(E + M\sqrt{\rho})},$$

and hence we have

$$0 < \frac{E}{M} - \sqrt{\rho} < \frac{\sqrt{\rho}}{M(E + M\sqrt{\rho})} = \frac{1}{M^2(E/(M\sqrt{\rho}) + 1)}.$$

Also, $0 < E/M - \sqrt{\rho}$ implies $E/(M\sqrt{\rho}) > 1$, and therefore

$$\left| \frac{E}{M} - \sqrt{\rho} \right| < \frac{1}{2M^2}.$$

By Theorem 2.1.14, E/M is a convergent in the continued fraction expansion of $\sqrt{\rho}$.
  If $N > 0$, we take $\sigma = N$, $\rho = D$, $E = s$, $M = t$, and the theorem holds in this case.
  If $N < 0$, then $t^2 - (1/D)s^2 = -N/D$, and we take $\sigma = -N/D$, $\rho = 1/D$, $E = t$, $M = s$. We find that t/s is a convergent in the expansion of $1/\sqrt{D}$. Then Theorem 2.1.15 shows that s/t is a convergent in the expansion of $\sqrt{D}$.   □

The following result is a corollary of Theorems 2.1.21, 3.3.1, and 3.3.3.

Theorem 3.3.4. *All positive solutions to $x^2 - Dy^2 = \pm 1$ are to be found among $x = h_n$, $y = k_n$, where h_n/k_n are the convergents of the expansion of $\sqrt{D}$. If r is the period of the expansion of $\sqrt{D}$, as in Theorem 2.1.21 and if r is even, then $x^2 - Dy^2 = -1$ has no solution, and all positive solutions to $x^2 - Dy^2 = 1$ are given by $x = h_{nr-1}$, $y = k_{nr-1}$ for $n = 1, 2, 3, \ldots$. On the other hand, if r is odd, then $x = h_{nr-1}$, $y = k_{nr-1}$ give all positive solutions to $x^2 - Dy^2 = -1$ for $n = 1, 3, 5, \ldots$, and all positive solutions to $x^2 - Dy^2 = 1$ for $n = 2, 4, 6, \ldots$.*

The sequences of pairs $(h_0, k_0), (h_1, k_1), \ldots$ will include all positive solutions to $x^2 - Dy^2 = 1$. Furthermore, $a_0 = [\sqrt{D}] > 0$, so the sequence $h_0, h_1, h_2, \ldots$ is strictly increasing. If we let (x_1, y_1) denote the first solution that appears, then for every other solution (x, y) we have $x > x_1$, and hence $y > y_1$ also. Having found this least positive solution by means of continued fractions, we can find all the remaining positive solutions by a simpler method, which is in fact similar to the second part of the proof of Theorem 3.2.1.   □

Following the same argument as in the last part of the proof in Theorem 3.2.1, we conclude that all nonnegative solutions are given by (x_n, y_n) for $n = 0, 1, 2, \ldots$, where x_n and y_n are the integers defined by $x_n + y_n\sqrt{D} = (x_1 + y_1\sqrt{D})^n$.

To illustrate the above method, we will consider the numerical example given by the equation $x^2 - 29y^2 = 1$. The expansion of $\sqrt{29}$ is $\sqrt{29} = \langle 5; \overline{2, 1, 1, 2, 10} \rangle$, so we have $n = 5$, an odd number. The first five convergents are $\frac{5}{1}, \frac{11}{2}, \frac{16}{3}, \frac{27}{5}$, $\frac{70}{13} = \frac{h_5}{k_5}$. But $x = h_5 = 70$, $y = k_5 = 13$ give $x^2 - 29y^2 = -1$. Hence, we must move on to the next period. The next period gives the convergents $\frac{727}{135}, \frac{1524}{283}$, $\frac{2251}{418}, \frac{3775}{701}, \frac{9801}{1820} = \frac{h_{10}}{k_{10}}$ and so by taking $x = h_{10} = 9801$, $y = k_{10} = 1820$, we obtain the smallest solution to our equation.

3.4 The General Solution by Quadratic Rings

The following proof uses results about quadratic rings introduced in Section 2.2. If D is a positive integer that is not a perfect square, consider the commutative quadratic ring $R = \{m + n\sqrt{D};\ m, n \in \mathbb{Z}\}$ endowed with the norm $N(\mu) = \mu \cdot \overline{\mu}$, where $\mu = a + b\sqrt{D}$ and $\overline{\mu} = a - b\sqrt{D}$.

For an element μ in R, $\mu \neq 0$, we will denote by $l(\mu)$ the vector in $\mathbb{R}^2$ defined by $l(\mu) = (\ln |\mu|, \ln |\overline{\mu}|)$.

The next result is fundamental for the method we are going to describe. For the proof we will use the approach given in [171] and [95].

Theorem 3.4.1. *In the ring R there exists a unit $\varepsilon_0 \neq \pm 1$ such that for any other unit ε in R the relation $\varepsilon = \pm \varepsilon_0^k$ holds for some integer k and some choice of signs $+$ and $-$.*

Proof. Let q be a real number such that $q > 2\sqrt{D}$. For all nonzero elements α in R with $|N(\alpha)| \leq q$, we denote by Y_α the set in $\mathbb{R}^2$ given by

$$Y_\alpha = \{(x, y) \in H :\ x \geq \ln |\alpha| \text{ and } y \geq \ln |\overline{\alpha}|\},$$

where H is the plane defined by the equation $x + y = \ln q$.

We will first prove that for all nonzero α in R the set Y_α is bounded in $\mathbb{R}^2$. Indeed, if $(x, y) \in Y_\alpha$, then $x \geq \ln |\alpha|$ and $y \geq \ln |\overline{\alpha}|$. Taking into account that $x + y = \ln q$ yields $x = \ln q - y \leq \ln q - \ln |\overline{\alpha}|$ and $y = \ln q - x \leq \ln q - \ln |\alpha|$, it follows that Y_α is contained into a rectangle in H. Moreover, if $|N(\alpha)| \leq q$, then Y_α is nonempty. Indeed, the inequality $|N(\alpha)| = |\alpha \cdot \overline{\alpha}| \leq q$ implies $\ln |\alpha| + \ln |\overline{\alpha}| \leq q$, hence $Y_\alpha \neq \emptyset$.

We will show now that for any unit ε in R the following equality holds:

$$Y_{\alpha\varepsilon} = Y_\alpha + l(\varepsilon).$$

This means that $x + y = \ln q$, $x \geq \ln |\alpha|$ and $y \geq \ln |\overline{\alpha}|$. Let

$$(x_1, y_1) = (x, y) + l(\varepsilon) = (x + \ln |\varepsilon|, y + \ln |\overline{\varepsilon}|).$$

Then

$$x_1 + y_1 = x + y + \ln |\varepsilon| + \ln |\overline{\varepsilon}| = x + y + \ln |\varepsilon \cdot \overline{\varepsilon}| = \ln q,$$

because $x + y = \ln q$ and $|\varepsilon \cdot \overline{\varepsilon}| = |N(\varepsilon)| = 1$. From Proposition 2.2.2, ε is a unit of R if and only if $N(\varepsilon) = \pm 1$. Also

$$x_1 = x + \ln |\varepsilon| \geq \ln |\alpha| + \ln |\varepsilon| = \ln |\alpha\varepsilon|$$

and

$$y_1 = y + \ln |\overline{\varepsilon}| \geq \ln |\overline{\alpha}| + \ln |\overline{\varepsilon}| = \ln |\overline{\alpha} \cdot \overline{\varepsilon}| = \ln |\overline{\alpha\varepsilon}|,$$

because from Proposition 2.2.4 the conjugate is multiplicative. This shows that $(x_1, y_1) \in Y_{\alpha \varepsilon}$, hence we have the inclusion

$$Y_\alpha + l(\varepsilon) \subseteq Y_{\alpha \varepsilon}.$$

For the converse inclusion consider $(x_1, y_1) \in Y_{\alpha \varepsilon}$. This means that $x_1 + y_1 = \ln q$ and

$$x_1 \geq \ln |\alpha \varepsilon|, \quad y_1 \geq \ln |\bar\alpha \cdot \bar\varepsilon| = \ln |\overline{\alpha \varepsilon}|.$$

Letting $x = x_1 - \ln |\varepsilon|$ and $y = y_1 - \ln |\bar\varepsilon|$ we have

$$x + y = \ln q,$$

$$x \geq \ln |\alpha \varepsilon| - \ln |\varepsilon| = \ln |\alpha|,$$

$$y \geq \ln |\overline{\alpha \varepsilon}| - \ln |\bar\varepsilon| = \ln |\bar\alpha|.$$

It follows that $(x, y) \in Y_\alpha$ and that $(x_1, y_1) = (x, y) + l(\varepsilon)$, i.e., $Y_{\alpha \varepsilon} \subseteq Y_\alpha + l(\varepsilon)$. Therefore, for any nonzero element α in R with $|N(\alpha)| \leq q$ and for any unit ε in R, we have $Y_{\alpha \varepsilon} = Y_\alpha + l(\varepsilon)$.

Now we will prove that

$$H \subseteq \bigcup_{\substack{|N(\alpha)| \leq q \\ \alpha \in R, \alpha \neq 0}} Y_\alpha.$$

For this, let $(x, y) \in H$ and let $x_1, y_1 \in \mathbb{R}_+^*$ such that $x = \ln x_1$ and $y = \ln y_1$. The equality $x + y = \ln q$ implies $x_1 y_1 = q$. Denote

$$X = [-x_1, x_1] \times [-y_1, y_2].$$

If λ is the Lebesgue measure in $\mathbb{R}^2$, then

$$\lambda(X) = 4x_1 y_1 = 4q > 4 \cdot 2\sqrt{D} = 4\lambda(T),$$

where $T = \{x(1, 1) + y(\sqrt{D}, -\sqrt{D}) : x, y \in [0, 1)\}$ is the fundamental parallelepiped associated with the complete lattice in $\mathbb{R}^2$, $\Lambda = \{m(1, 1) + n(\sqrt{D}, -\sqrt{D}) : m, n \in \mathbb{Z}\}$. The lattice Λ is complete because the vectors $(1, 1)$ and $(\sqrt{D}, -\sqrt{D})$ are linearly independent over $\mathbb{R}$. It is known that $\lambda(T) = |\det A_D| = 2\sqrt{D}$, where A_D is the matrix

$$A_D = \begin{pmatrix} 1 & 1 \\ \sqrt{D} & -\sqrt{D} \end{pmatrix}.$$

Using the Minkowski's Fundamental Theorem (see [165]), it follows that there exist integers m and n such that

$$(m, n) \neq (0, 0) \text{ and } m(1, 1) + n(\sqrt{D}, -\sqrt{D}) \in X \cap \Lambda.$$

From the definition of the set X we obtain

$$|m + n\sqrt{D}| \leq x_1 \text{ and } |m - n\sqrt{D}| \leq y_1.$$

Setting $\alpha = m + n\sqrt{D}$ and taking into account that $(m, n) \neq (0, 0)$ yields that α is a nonzero element of the ring R and that

$$|N(\alpha)| = |\alpha\bar{\alpha}| = |\alpha||\bar{\alpha}| = |m + n\sqrt{D}||m - n\sqrt{D}| \leq x_1 y_1 = q.$$

Because

$$x = \ln x_1 \geq \ln|m + n\sqrt{D}| = \ln|\alpha| \text{ and } y = \ln y_1 \geq \ln|m - n\sqrt{D}| = \ln|\bar{\alpha}|,$$

it follows that $(x, y) \in Y_\alpha$, i.e., the inclusion $H \subseteq \cup Y_\alpha$ is proved. By using Theorem 2.2.3 we deduce the existence of a finite number of elements $\alpha_1, \alpha_2, \ldots, \alpha_r \in R$ with the property that each α in R with $|N(\alpha)| \leq q$ is divisibility associated with one of the elements $\alpha_1, \alpha_2, \ldots, \alpha_r$.

The sets Y_{α_i}, $i = 1, 2, \ldots, r$ are bounded, hence the set $Y = \bigcup_{i=1}^{r} Y_{\alpha_i}$ is also bounded in $\mathbb{R}^2$. Let $(x, y) \in H$. Using the above considerations, it follows that there exists a nonzero element α in R such that $|N(\alpha)| \leq q$ and that $(x, y) \in Y_\alpha$. By the choice of elements $\alpha_1, \alpha_2, \ldots, \alpha_r$, there exists $i \in \{1, 2, \ldots, r\}$ such that $\alpha = \varepsilon \alpha_i$, where ε is unit in the ring R. Hence

$$(x, y) \in Y_\alpha = Y_{\alpha_i \varepsilon} = Y_{\alpha_i} + l(\varepsilon),$$

and so $H \subseteq Y + L$, where

$$L = \{l(\varepsilon) : \varepsilon \text{ unit in } R\}.$$

It is clear that $(0, 0) \in L$, because $(0, 0) = l(1)$, and that $(L, +)$ is a subgroup of the commutative group $(\mathbb{R}^2, +)$. Since the set Y is bounded, and the set H is not, it follows that L is an infinite set, in particular $L \neq \{(0, 0)\}$. Assume there is a sequence $(\varepsilon_n)_{n \geq 1}$ of units in R such that $\lim_{n \to \infty} l(\varepsilon_n) = (0, 0)$ and that $\varepsilon_n \neq \pm 1$ for all positive integers n. This shows that $\lim_{n \to \infty} |\varepsilon_n| = \lim_{n \to \infty} |\bar{\varepsilon}_n| = 1$, and so $\lim_{n \to \infty} \max\{|\varepsilon_n|, |\bar{\varepsilon}_n|\} = 1$. It is not difficult to see that either $|\varepsilon_n|$ or $|\bar{\varepsilon}_n|$ has the form $m + m'\sqrt{D}$ for some nonnegative integers m and m'.

For $n \geq 1$, $\max\{|\varepsilon_n|, |\bar{\varepsilon}_n|\} \geq \sqrt{D} \geq 2$, so $D \geq 2$. For $n = 0$, taking into account that $\varepsilon_n \neq \pm 1$ for all n, yields $\max\{|\varepsilon_n|, |\bar{\varepsilon}_n|\} \geq m \geq 2$. In both cases, $\max\{|\varepsilon_n|, |\bar{\varepsilon}_n|\} \geq \sqrt{2}$, so $\lim\limits_{n \to \infty} \max\{|\varepsilon_n|, |\bar{\varepsilon}_n|\} \neq 1$ and $\lim\limits_{n \to \infty} l(\varepsilon_n) \neq (0,0)$. From all of the above, it follows that there is a unit $\varepsilon_0 \neq \pm 1$ in R such that

$$\|l(\varepsilon_0)\| = \min\{\|l(\varepsilon)\| : \varepsilon \text{ is a unit in } R, \ \varepsilon \neq \pm 1\},$$

where $\| \cdot \|$ denotes the well-known Euclidean norm in $\mathbb{R}^2$. We have used above that $l(\varepsilon) = (0,0)$ if and only if $\varepsilon = \pm 1$. In particular, it follows that $\|l(\varepsilon_0)\| > 0$. Replacing, if necessary, ε_0 by $\pm\bar{\varepsilon}_0$ or by $-\varepsilon_0$, and taking into account that

$$\|l(\varepsilon_0)\| = \|l(-\varepsilon_0)\| = \|l(\bar{\varepsilon}_0)\| = \|l(-\bar{\varepsilon}_0\|,$$

one can assume that $\varepsilon_0 = m + m'\sqrt{D}$, where m, m' are nonnegative integers such that $(m, m') \neq (1, 0)$. This means that $\varepsilon_0 > 1$. Such a unit ε_0 is called the fundamental unit of the ring R. Since $\ln |\varepsilon| + \ln |\bar{\varepsilon}| = \ln 1 = 0$, for all units ε in R, the following relation holds: $L \subseteq \{(x,y) \in \mathbb{R}^2 : x + y = 0\}$. If $l(\varepsilon_0) = (\alpha, -\alpha)$, where $\alpha = \ln |\varepsilon_0| = \ln \varepsilon_0 > 0$ and $l(\varepsilon) = (\beta, -\beta)$ is another element of the set L with $\beta > 0$, let k be a positive integer such that $k\alpha \leq \beta \leq (k+1)\alpha$ (we have $\|l(\varepsilon)\| = \beta\sqrt{2} \geq \|l(\varepsilon_0)\| = \alpha\sqrt{2}$, hence $\beta \geq \alpha$ and $k \geq 1$). Let $\varepsilon_1 \in R, \varepsilon_1 = \varepsilon \cdot \varepsilon_0^{-k}$. Then

$$l(\varepsilon_1) = l(\varepsilon) - kl(\varepsilon_0) = (\beta - k\alpha, -\beta + k\alpha).$$

If $\beta - k\alpha > 0$, then $\varepsilon_1 \neq \pm 1$ and

$$\|l(\varepsilon_1)\| = \sqrt{2}(\beta - k\alpha) < \sqrt{2}((k+1)\alpha - k\alpha) = \sqrt{2} \cdot \alpha = \|l(\varepsilon_0)\|,$$

in contradiction with the choice of ε_0. Therefore $\beta = k\alpha$, $l(\varepsilon_1) = 0$, which implies the equality $\varepsilon = \pm\varepsilon_0^k$. Note that if $l(\varepsilon) = (\beta, -\beta)$ and $\beta < 0$, then the same argument above for $\bar{\varepsilon}$ shows that there exists a nonnegative integer k with the property $\bar{\varepsilon} = \pm\varepsilon_0^k$.

From all the considerations above it follows that all units in the ring R are of the form $\pm\varepsilon_0^k$ for some integer k. $\qquad\square$

Theorem 3.4.1 facilitates finding all positive integer solutions to the Pell's equation $x^2 - Dy^2 = 1$. In this respect, consider a solution (u, v) and denote $\varepsilon = u + v\sqrt{D}$. Then $N(\varepsilon) = u^2 - Dv^2 = 1$, so ε is a unit in the ring R. Applying the result in the Theorem 3.4.1, it follows that $\varepsilon = \pm\varepsilon_0^k$, for some k and for some choice of signs $+$ and $-$. In addition, if we assume $(u, v) \neq (1, 0)$, then $\varepsilon > 1$. Taking into consideration that $\varepsilon_0 > 1$ and that $\varepsilon = \pm\varepsilon_0^k > 1$, we see that the integer k is positive and that we must to choose the sign $+$. Therefore, $\varepsilon = \varepsilon_0^k$, where k is a positive integer. Moreover, if $N(\varepsilon_0) = -1$, then one needs the necessary condition k even (indeed, $1 = N(\varepsilon) = N(\varepsilon_0)^k = (-1)^k$ in the case $N(\varepsilon_0) = -1$).

The general solution to Pell's equation $x^2 - Dy^2 = 1$ could be also written recursively as follows:

$$x_0 = 1, \ y_0 = 0$$
$$x_1 = m, \ y_1 = n \text{ if } N(\varepsilon_0) = 1, \ \varepsilon_0 = m + n\sqrt{D}$$
$$x_1 = m^2 + Dn^2, \ y_1 = 2mn \text{ if } N(\varepsilon_0) = -1$$

$$\begin{cases} x_{k+1} = mx_k + Dny_k \\ y_{k+1} = nx_k + my_k \end{cases}, \text{ if } N(\varepsilon_0) = 1$$

$$\begin{cases} x_{k+1} = (m^2 + Dn^2)x_k + 2Dmny_k \\ y_{k+1} = 2mnx_k + (m^2 + Dn^2)y_k \end{cases}, \text{ if } N(\varepsilon_0) = -1.$$

3.5   The Equation $ax^2 - by^2 = 1$

In the present section we will study the more general equation

$$ax^2 - by^2 = 1, \tag{3.5.1}$$

where a and b are positive integers. Taking into account the considerations in Section 3.1 we have $\Delta = 4ab > 0$, hence (3.5.1) can be reduced to a Pell's equation. In the paper [144] is given a continued fraction approach.

We will use the results in [13, 14] and [15].

Proposition 3.5.1. *If* $ab = k^2$, *where* k *is an integer greater than 1, then the equation (3.5.1) does not have solutions in positive integers.*

Proof. Assume that (3.5.1) has a solution (x, y), where x, y are positive integers. Then $ax^2 - by^2 = 1$, and clearly a and b are relatively prime. From the condition $ab = k^2$ it follows that $a = k_1^2$ and $b = k_2^2$ for some positive integer k_1 and k_2. The relation $k_1^2 x^2 - k_2^2 y^2 = 1$ can be written as $(k_1 x - k_2 y)(k_1 x + k_2 y) = 1$. It follows that $1 < k_1 x + k_2 y = k_1 x - k_2 y = 1$, a contradiction. □

We will call *Pell's resolvent* of (3.5.1) the equation

$$u^2 - abv^2 = 1. \tag{3.5.2}$$

Theorem 3.5.2. *Suppose that the equation (3.5.1) has solutions in positive integers and let* (x_0, y_0) *be its smallest solution. The general solution to (3.5.1) is* $(x_n, y_n)_{n \geq 0}$, *where*

$$x_n = x_0 u_n + by_0 v_n, \quad y_n = y_0 u_n + ax_0 v_n, \tag{3.5.3}$$

and $(u_n, v_n)_{n \geq 0}$ *is the general solution to Pell's resolvent (3.5.2).*

Proof. We will prove first that $(x_n, y_n)_{n \geq 0}$ is a solution to the equation (3.5.1). Indeed,

$$ax_n^2 - by_n^2 = a(x_0 u_n + by_0 v_n)^2 - b(y_0 u_n + ax_0 v_n)^2 =$$
$$= (ax_0^2 - by_0^2)(u_n^2 - abv_n^2) = 1 \cdot 1 = 1.$$

Conversely, let (x, y) be a solution to the equation (3.5.1). Then (u, v), where $u = ax_0 x - by_0 y$ and $v = y_0 x - x_0 y$, is a solution to Pell's resolvent (3.5.2). Solving the above system of linear equations with unknowns x and y yields $x = x_0 u + by_0 v$ and $y = y_0 u + ax_0 v$, i.e., (x, y) has the form (3.5.3).   □

Remarks. 1) A simple algebraic computation yields the following relation between the fundamental solution (u_1, v_1) to Pell's resolvent and the smallest solution (x_0, y_0) to equation (3.5.1): $u_1 \pm v_1 \sqrt{ab} = (x_0 \sqrt{a} \pm y_0 \sqrt{b})^2$, where the signs + and − correspond.

2) Using formulas (3.2.6), from (3.5.3) it follows that

$$x_n = \frac{1}{2}\left(x_0 + \frac{y_0}{a}\sqrt{ab}\right)\left(u_1 + v_1\sqrt{ab}\right)^n + \frac{1}{2}\left(x_0 - \frac{y_0}{a}\sqrt{ab}\right)\left(u_1 - v_1\sqrt{ab}\right)^n$$
$$y_n = \frac{1}{2}\left(y_0 + \frac{x_0}{b}\sqrt{ab}\right)\left(u_1 + v_1\sqrt{ab}\right)^n + \frac{1}{2}\left(y_0 - \frac{x_0}{b}\sqrt{ab}\right)\left(u_1 - v_1\sqrt{ab}\right)^n.$$

Taking into account Remark 1, the above formulas can be written as

$$x_n = \frac{1}{2\sqrt{a}}\left[\left(x_0\sqrt{a} + y_0\sqrt{b}\right)^{2n+1} + \left(x_0\sqrt{a} - y_0\sqrt{b}\right)^{2n+1}\right]$$
$$y_n = \frac{1}{2\sqrt{b}}\left[\left(x_0\sqrt{a} + y_0\sqrt{b}\right)^{2n+1} - \left(x_0\sqrt{a} - y_0\sqrt{b}\right)^{2n+1}\right].$$

This last form of solutions appears in [219] but the method given there is much more complicated.

3) The general solution (3.5.3) can be written in the following matrix form

$$\begin{pmatrix} x_n \\ y_n \end{pmatrix} = \begin{pmatrix} x_0 & by_0 \\ y_0 & ax_0 \end{pmatrix}\begin{pmatrix} u_n \\ v_n \end{pmatrix} = \begin{pmatrix} x_0 & by_0 \\ y_0 & ax_0 \end{pmatrix}\begin{pmatrix} u_1 & abv_1 \\ v_1 & u_1 \end{pmatrix}^n \begin{pmatrix} u_0 \\ v_0 \end{pmatrix}.$$

To illustrate the above method, let us consider the following equation: $6x^2 - 5y^2 = 1$. Its smallest solution is $(x_0, y_0) = (1, 1)$. The Pell's resolvent is $u^2 - 30v^2 = 1$, whose fundamental solution is $(11, 2)$. The general solution to the equation considered is $x_n = u_n + 5v_n$, $y_n = u_n + 6v_n$, $n = 0, 1, \ldots$ where $(u_n, v_n)_{n \geq 0}$ is the general solution to Pell's resolvent, i.e., $u_{n+1} = 11u_n + 60v_n$, $v_{n+1} = 2u_n + 11v_n$, $n = 0, 1, \ldots$ with $u_0 = 11$, $v_0 = 2$.

A closed form for these solutions can be found by using the above Remark 2. We obtain

$$x_n = \frac{1}{2\sqrt{6}} \left[\left(\sqrt{6} + \sqrt{5} \right)^{2n+1} + \left(\sqrt{6} - \sqrt{5} \right)^{2n+1} \right]$$

$$y_n = \frac{1}{2\sqrt{5}} \left[\left(\sqrt{6} + \sqrt{5} \right)^{2n+1} - \left(\sqrt{6} - \sqrt{5} \right)^{2n+1} \right].$$

Remarks. 1) In the paper [84] is given a nice survey concerning the history and various approaches in solving the equation (3.5.1).

2) The next result is due in [152]: If $1 < a < b$ are integers such that ab is square-free, then at most one of the two equations

$$ax^2 - by^2 = \pm 1 \tag{3.5.4}$$

is solvable.

3) In Example 3, page 140 of [22], it is shown that if $a, b \geq 1$ are integers such that ab is not a perfect square and both equations (3.5.4) are solvable, then $a = 1$ or $b = 1$.

3.6 The Negative Pell Equation and the Pell–Stevenhagen Constants

While the Pell's equation $x^2 - Dy^2 = 1$ is always solvable if the positive integer D is not a perfect square, as we have proven in the previous sections, the equation

$$x^2 - Dy^2 = -1 \tag{3.6.1}$$

is solvable only for certain values of D.

We have seen in Theorem 3.3.4 that if r is the period of the expansion of $\sqrt{D}$ in continued fractions, then, if r is even, the equation (3.6.1) has no solution. If r is odd, then $x = h_{nr-1}$ and $y = k_{nr-1}$ give all positive solutions to (3.6.1) for $n = 1, 3, 5, \ldots$.

Next, we will write the solutions to the equation (3.6.1) by using our method in Section 3.5.

The equation (3.6.1) is known as the *negative Pell's equation*. From the Theorem 3.5.2 the following theorem follows:

Theorem 3.6.1. *Suppose that the equation (3.6.1) has solutions in positive integers and let (x_0, y_0) be its smallest solution. The general solution to (3.6.1) is given by $(x_n, y_n)_{n \geq 0}$ where*

$$x_n = x_0 u_n + Dy_0 v_n, \quad y_n = y_0 u_n + x_0 v_n \tag{3.6.2}$$

and $(u_n, v_n)_{n \geq 0}$ is the general solution to Pell's equation $u^2 - Dv^2 = 1$.

Remarks. 1) Between (x_0, y_0) and (u_1, v_1) there is the following important connection:

$$u_1 \pm v_1 \sqrt{D} = \left(x_0 \pm y_0 \sqrt{D} \right)^2,$$

where the signs + and − correspond.

2) By using formulas (3.6.2) we obtain the solutions to the negative Pell's equation in explicit form:

$$x_n = \frac{1}{2} \left(x_0 + y_0 \sqrt{D} \right) \left(u_1 + v_1 \sqrt{D} \right)^n + \frac{1}{2} \left(x_0 - y_0 \sqrt{D} \right) \left(u_1 - v_1 \sqrt{D} \right)^n$$

$$y_n = \frac{1}{2} \left(y_0 + \frac{x_0}{D} \sqrt{D} \right) \left(u_1 + v_1 \sqrt{D} \right)^n + \frac{1}{2} \left(y_0 - \frac{x_0}{D} \sqrt{D} \right) \left(u_1 - v_1 \sqrt{D} \right)^n.$$
(3.6.3)

Formulas (3.6.3) can be also written as

$$x_n = \frac{1}{2} \left[\left(x_0 + y_0 \sqrt{D} \right)^{2n+1} + \left(x_0 - y_0 \sqrt{D} \right)^{2n+1} \right]$$
(3.6.4)

$$y_n = \frac{1}{2\sqrt{D}} \left[\left(x_0 + y_0 \sqrt{D} \right)^{2n+1} - \left(x_0 - y_0 \sqrt{D} \right)^{2n+1} \right]$$

3) The general solution (3.6.2) can be expressed in the following matrix form

$$\begin{pmatrix} x_n \\ y_n \end{pmatrix} = \begin{pmatrix} x_0 & Dy_0 \\ y_0 & x_0 \end{pmatrix} \begin{pmatrix} u_n \\ v_n \end{pmatrix} = \begin{pmatrix} x_0 & Dy_0 \\ y_0 & x_0 \end{pmatrix} \begin{pmatrix} u_1 & Dv_1 \\ v_1 & u_1 \end{pmatrix}^n \begin{pmatrix} u_0 \\ v_0 \end{pmatrix}.$$

The following result points out an important class of solvable negative Pell's equations. The proof is adapted from [151].

Theorem 3.6.2. *Let p be a prime ≥ 3. The negative Pell's equation*

$$x^2 - py^2 = -1$$

is solvable in positive integers if and only if $p \equiv 1 \pmod 4$.

Proof. First suppose that the equation is solvable. Then there are positive integers u, v such that $u^2 - pv^2 = -1$. So, $u^2 - (-1) = pv^2$, implying $\left(\dfrac{-1}{p} \right) = 1$, where $\left(\dfrac{a}{p} \right)$ denotes the Legendre symbol. According to Theorem 4.3.1.5) in [22, pp. 178–179] we have $\left(\dfrac{-1}{p} \right) = (-1)^{\frac{p-1}{2}}$, hence $p \equiv 1 \pmod 4$.

Let (u_0, v_0) be the fundamental solution to the Pell's resolvent $u^2 - pv^2 = 1$. Then $u_0^2 - 1 = pv_0^2$, and u_0 cannot be even, for in this case we should have $-1 \equiv p \pmod 4$. Hence u_0 is odd and the numbers $u_0 - 1$ and $u_0 + 1$ have the greatest common divisor 2. Therefore $u_0 \pm 1 = 2\alpha^2$ and $u_0 \mp 1 = 2p\beta^2$, where α and β are positive integers such that $v_0 = 2\alpha\beta$. By elimination of u_0 we get $\pm 1 = \alpha^2 - p\beta^2$. Since $\beta < v_0$, we cannot have the upper sign. Thus the lower sign must be taken and the theorem is proved. $\qquad\square$

Remarks. 1) To give an example of an unsolvable negative Pell's equation we will show that the equation $x^2 - 34y^2 = -1$ has no solution. The fundamental solution of Pell's resolvent is $(35, 6)$. If the equation $x^2 - 34y^2 = -1$ were solvable and had the fundamental solution (x_0, y_0), then by Theorem 3.3.4 we would have $x_0^2 + 34y_0^2 = 35$ and $2x_0y_0 = 6$. But this system of equations has no solutions in positive integers and thus the equation $x^2 - 34y^2 = -1$ is not solvable.

2) In the paper [58] is proved that the proportion of square-free D divisible by k primes of the form $4m + 1$ for which the negative Pell equation is solvable is at least 40 %.

    The following short and completely elementary criterion concerning the solvability of the negative Pell equation was obtained in the paper [146].

Theorem 3.6.3. *If $D \equiv 1, 2 \pmod 4$ is a non-square integer, then there is a solution to (3.6.1) if and only if $u_1 \equiv -1 \pmod{2D}$, where (u_1, v_1) is the fundamental solution to the Pell equation $u^2 - Dv^2 = 1$.*

Proof. If (3.6.1) is solvable with the smallest solution (x_0, y_0), then we have $u_1 = x_0^2 + Dy_0^2 = -1 + 2Dy_0^2 \equiv -1 \pmod{2D}$ (see Remark 1) after Theorem 3.6.1).

    Conversely, assume that the fundamental solution (u_1, v_1) to $u^2 - Dv^2 = 1$ satisfies $u_1 \equiv -1 \pmod{2D}$. It follows that $u_1 = -1 + 2Dk$, for some positive integer k. We have $(-1 + 2Dk)^2 - Dv_1^2 = 1$, which gives $Dk^2 - k - v_1'^2 = 0$, where $v_1 = 2v_1'$. Therefore,

$$k(Dk - 1) = v_1'^2,$$

from which it follows that $k = r^2$ and $Dk - 1 = s^2$ as $\gcd(k, Dk - 1) = 1$. Thus $Dk - 1 = Dr^2 - 1 = s^2$ which gives $s^2 - Dr^2 = -1$, hence the negative Pell equation is solvable. $\qquad\square$

Remark. In [147] is explored the central norm in the simple continued fraction expression of $\sqrt{D}$, where $D \geq 2$ is not a perfect square. The obtained results are used by the authors in the study of solvability of the negative Pell's equation.

    In what follows we will present a result concerning the negative Pell's equation based on our paper [18]. We begin with a representation theorem of the Fibonacci sequence that will turn to be useful in the proof of our result.

    We consider the Diophantine equation

$$x^2 + y^2 + 1 = xyz. \tag{3.6.5}$$

First we will establish the necessary condition of solvability for equation (3.6.5) and then we will determine all its solutions in terms of the well-known Fibonacci sequence $(F_m)_{m \geq 1}$.

Theorem 3.6.4. *The equation (3.6.5) is solvable if and only if $z = 3$. In this case all of its solutions (x, y) are given by*

$$(1, 1), \ (F_{2n+1}, F_{2n-1}), \ (F_{2n-1}, F_{2n+1}), \ n \geq 1.$$

Proof. Let (x, y, z) be a solution with $z \neq 3$. Then $x \neq y$, for otherwise $x^2(z-2) = 1$, which is impossible, since $z - 2 \neq 1$. We have

$$0 = x^2 + y^2 + 1 - xyz = (x - yz)^2 + y^2 + 1 + xyz - y^2 z^2$$
$$= (yz - x)^2 + y^2 + 1 - (yz - x)yz$$

hence $(yz - x, y, z)$ is also a solution, since $x(yz - x) = xyz - x^2 = y^2 + 1 > 0$ implies $yz - x > 0$. Note that if $x > y$, then $x^2 > y^2 + 1 = x(yz - x)$. Hence $x > yz - x$, which shows that the newly obtained solution is smaller than the initial solution, in the sense that $x + y > (yz - x) + y$. However, under the assumption that $x \neq y$, this procedure can be continued indefinitely, which is impossible, since in the process we construct a decreasing sequence of positive integers, a contradiction. This contradiction shows that there are no solutions if $z \neq 3$.

Clearly, $(1,1)$ is a solution to the equation

$$x^2 + y^2 + 1 = 3xy.$$

Let (a, b), $a > b$, be another solution. Then $b^2 + (3b - a)^2 + 1 = 3b(3b - a)$, so $(b, 3b - a)$ is also a solution. From

$$(a - b)(a - 2b) = a^2 - 3ab + 2b^2 = b^2 - 1 > 0$$

it follows that $a > 2b$, hence $3b - a < b$. So the new solution has a smaller b.

It follows that we reach a solution with $b = 1$, hence with $a^2 + 2 = 3a$, in which case $a = 1$ or $a = 2$. It follows that all solutions are obtained from $(a_1, b_1) = (1, 1)$ by the recursion

$$(a_{n+1}, b_{n+1}) = (b_n, 3b_n - a_n).$$

The sequences $(a_n)_{n \geq 1}$ and $(b_n)_{n \geq 1}$ satisfy the same recursion: $x_{n+1} = 3x_n - x_{n-1}$, $x_1 = 1$, $x_2 = 2$. This recursion characterizes the Fibonacci numbers of odd index. Therefore, $(a_n, b_n) = (F_{2n+1}, F_{2n-1})$, $n \geq 1$.

The solutions are $(1, 1)$, (F_{2n+1}, F_{2n-1}), (F_{2n-1}, F_{2n+1}), for $n \geq 1$.   □

The following result points out a family of unsolvable negative Pell's equations [199]:

Theorem 3.6.5. *Let k be an integer greater than 2. The equation*

$$x^2 - (k^2 - 4)y^2 = -1 \tag{3.6.6}$$

is solvable if and only if k = 3.

Proof. We will show that the equation

$$u^2 - (k^2 - 4)v^2 = -4 \tag{3.6.7}$$

is not solvable if $k \neq 3$. Assume the contrary, and let (u, v) be a solution to (3.6.7). Then u and kv have the same parity. Consider $x = \dfrac{u + kv}{2}$. Then $u = 2x - kv$ and (3.6.7) becomes

$$x^2 + v^2 + 1 = xvk.$$

Since $k \neq 3$, this contradicts the result in Theorem 3.6.4.

Assume now that for $k \neq 3$, the negative Pell's equation (3.6.6) has a solution (x, y). Multiplying both sides by 4 yields

$$(2x)^2 - (k^2 - 4)(2y)^2 = -4,$$

contradicting the above result concerning equation (3.6.7).

When $k = 3$ equation (3.6.6) becomes

$$x^2 - 5y^2 = -1. \tag{3.6.8}$$

The smallest solution to (3.6.8) is (2,1). From formulas (3.6.3) it follows that all solutions to (3.6.8) are given by $(x_n, y_n)_{n \geq 0}$, where

$$x_n = \frac{1}{2}\left[\left(2 + \sqrt{5}\right)^{2n+1} + \left(2 - \sqrt{5}\right)^{2n+1}\right]$$

$$y_n = \frac{1}{2\sqrt{5}}\left[\left(2 + \sqrt{5}\right)^{2n+1} - \left(2 - \sqrt{5}\right)^{2n+1}\right]$$

Remark. A complicated method for proving the result in Theorem 3.6.4 was given in [169].

The problem of determining those D for which the negative Pell's equation (3.6.1) is solvable in positive integers has a long history. We mentioned at the beginning of this section that if D is a positive nonsquare the solvability or unsolvability of (3.6.1) can be determined by expanding $\sqrt{D}$ as an ordinary continued fraction

$$\sqrt{D} = \langle a_0; \overline{a_1, \ldots, a_r} \rangle.$$

Then (3.6.1) is solvable or not according to whether r is odd or even. If r is odd, then

$$\frac{x_0}{y_0} = \langle a_0, a_1, \ldots, a_{r-1} \rangle$$

is the fundamental solution of (3.6.1).

A second approach to this problem involves using generalized residue symbol criteria derived from D to determine conditions on D which guarantee that (3.6.1) is solvable or unsolvable. This approach was initiated by Legendre in 1785. He proved that if D is a prime congruent to $1 \pmod 4$, then (3.6.1) is solvable (see Theorem 3.6.2), while if a prime p congruent to $3 \pmod 4$ divides the squarefree part of D, then (3.6.1) is unsolvable. Dirichlet observed that $D = pq$ with $p \equiv q \equiv 1 \pmod 4$ and $(p/q)_4 = (q/p)_4 = -1$, then (3.6.1) is solvable. For $D = p_1 \ldots p_N$ in [210] are given quadratic residue criteria among p which when they held would guarantee that (3.6.1) is solvable. In the paper [195] applied methods of class field theory were used to show that in the case $D = pq$ with $p \equiv q \equiv 1 \pmod 4$ equation (3.6.1) is unsolvable when $(p/q)_4 \neq (q/p)_4$, while in the case $(p/q)_4 = (q/p)_4 = 1$ the equation (3.6.1) is sometimes solvable and sometimes not. In the papers [195] and [181] it is proved that these residue symbol criteria were related to the structure of the 2-Sylow subgroup of an appropriate ring class group $\mathbb{Q}[\sqrt{D}]$. In [180, 181] is introduced a "conditional Artin symbol" defined in terms of generators of certain class fields, by means of which it is given a set of necessary and sufficient conditions for (3.6.1) to be solvable. In [153] it is acknowledged that the problem of determining those D for which (3.6.1) is solvable is still open, presumably due to the nonexplicit character of conditions in [180] and [181]. Explicit residue symbol conditions for special types of D are still being found, e.g., [99] and [177].

The residue symbol approach can be extended to yield an algorithm determining the solvability of negative Pell's equation whose main bottleneck is finding a factorization of D. In the paper [108] it is proved that there is an algorithm when given a positive integer D together with (i) a complete prime factorization of D and (ii) a quadratic nonresidue n_i for each prime p_i dividing D, determines whether the equation (3.6.1) is solvable in positive integers or not, and which always terminates in $O((\ln D)^5 (\ln \ln D)(\ln \ln \ln D))$ elementary operations.

In the paper [81] it is shown that for a nonsquare positive integer D, the negative Pell equation (3.6.1) is solvable if and only if there exist a primitive Pythagorean triple (A, B, C) and positive integers a, b such that $D = a^2 + b^2$ and $aA - bB = \pm 1$. It is also possible to describe a method to generate families of such integers D stemming from the solutions to the linear equation $aA - bB = \pm 1$.

If p is a prime such that $2p = a^4 + s^2$, where $a^2 \equiv \pm s \equiv 9 \pmod{16}$, then the negative Pell's equation $x^2 - 2py^2 = -1$ has no solution [44]. If $D = 2p$, where $p = c^2 + 8D^2$ and D is odd, then the equation (3.6.1) has no solutions [69]. If $p = c^2 + qD^2$ and D is odd when $p \equiv 1 \pmod 4$ and $(p/q) = 1$, then the negative Pell's equation $t^2 - pqu^2 = -1$ has no solutions [109].

While the set of positive integers D for which the Pell's equation is solvable is well known (it is the set of all nonsquare positive integers), the set $\mathcal{D}$ of all positive integers D for which the negative Pell's equation is solvable is far from being known. Only recently has progress been made in the study of the set $\mathcal{D}$. We will mention a few results and some open problems concerning the set $\mathcal{D}$.

Without loss of generality one can, however, assume that D is square-free. Moreover, (3.6.1) can have solutions only if D has no prime divisors $\equiv 3 \pmod 4$. Consider the case in which $D = p_1' p_2' \ldots p_r'$, $p_1' < p_2' < \cdots < p_r'$ and $p_k' \equiv 1 \pmod 4$. In 1834, G. L. Dirichlet had shown that (3.6.1) has solutions when $r = 1$ (see Theorem 3.6.2) and also when $r = 2$ provided $(p_1'/p_2') = -1$. He had even considered the case when $r = 3$. In [155] it is shown that (3.6.1) has solutions for all odd r's, provided $(p_i'/p_j') = -1$ for each $i < j \leq r$.

Define the *Pell constant* (see [73, pp. 119–120])

$$P = 1 - \prod_{\substack{j \geq 1 \\ j \, odd}} \left(1 - \frac{1}{2^j}\right) = 0,5805775582\ldots$$

needed in what follows. The constant P is irrational [206] but only conjectured to be transcendental. Define also the function ψ by

$$\psi(p) = \frac{2 + (1 + 2^{1-\nu_p})p}{2(p+1)}$$

where ν_p is the exponent of 2 into factorization in $p - 1$.

For any set S of positive integers, let $f_S(n)$ denote the number of elements in S not exceeding n. In [43, 206, 207] several conjectures regarding the distribution of $\mathcal{D}$ were formulated. For example, it was conjectured that the counting function $f_\mathcal{D}$ satisfies the following relation [207]:

$$\lim_{n \to \infty} \frac{\sqrt{\ln n}}{n} f_\mathcal{D}(n) = \frac{3P}{2\pi} \prod_{\substack{p \, prime \\ p \equiv 1 (\mathrm{mod}\ 4)}} \left(1 + \frac{\psi(p)}{p^2 - 1}\right) \left(1 - \frac{1}{p^2}\right)^{1/2} = 0,28136\ldots$$

Let U be the set of positive integers not divisible by 4, and let V be the set of positive integers not divisible by any prime congruent to 3 modulo 4. Clearly, $\mathcal{D}$ is a subset of $U \cap V$, and $U \cap V$ is the set of positive integers that can be written as a sum of two coprime squares. By the above conjectured and by a coprimality result given in [182], the density of $\mathcal{D}$ inside $U \cap V$ is [207]:

$$\lim_{n \to \infty} \frac{f_\mathcal{D}(n)}{f_{U \cap V}(n)} = P \prod_{\substack{p \, prime \\ p \equiv 1 (\mathrm{mod}\ 4)}} \left(1 + \frac{\psi(p)}{p^2 - 1}\right) \left(1 - \frac{1}{p^2}\right) = 0,57339\ldots$$

Here is another conjecture involving the Pell constant. Let W be the set of squarefree integers, that is, integers which are divisible by no square exceeding 1. In [206] it is conjectured that

$$\lim_{n\to\infty} \frac{\sqrt{\ln n}}{n} f_{\mathcal{D}\cap W}(n) = \frac{6}{\pi^2} PK = 0,2697318462\ldots$$

where K is the Landau–Ramanujan constant. Clearly, U is a subset of W, and $V \cap W$ is the set of positive squarefree integers that can be written as a sum of two (coprime) squares. By the second conjectured limit and by a squarefree result one obtains that the density of $\mathcal{D} \cap W$ inside $V \cap W$ is [43]:

$$\lim_{n\to\infty} \frac{f_{\mathcal{D}\cap W}(n)}{f_{V\cap W}(n)} = P = 0,5805775582\ldots$$

An interesting connection to continued fractions is given in [207]: an integer $D > 1$ is in $\mathcal{D}$ if and only if $\sqrt{D}$ is irrational and has a regular continued fraction expansion with odd period length (see also Theorem 3.3.4).

If $D > 1$ is a squarefree integer with no prime factor p, $p \equiv 3 \pmod 4$, with exactly n prime factors, and if $\mathcal{D}_n(X)$ denotes the set of those $D \le X$, in the paper [58] the authors study the distribution of such D which lie in $\mathcal{D}$. An explicit number λ_n is given such that

$$\liminf_{X\to\infty} \frac{\#(\mathcal{D}_n(X) \cap \mathcal{D})}{\#\mathcal{D}_n(X)} \ge \lambda_n.$$

Moreover, it is conjectured that

$$\lim_{X\to\infty} \frac{\#(\tilde{\mathcal{D}}(X) \cap \mathcal{D})}{\#\tilde{\mathcal{D}}(X)} \ge \lambda_\infty$$

where $\tilde{\mathcal{D}}(X) = \bigcup_{n=1}^{\infty} \mathcal{D}_n(X)$, and $\lambda_\infty = \lim_{n\to\infty} \lambda_n = 0,419\ldots$.

Chapter 4
General Pell's Equation

This chapter gives the general theory and useful algorithms to find positive integer solutions (x, y) to general Pell's equation (4.1.1), where D is a nonsquare positive integer, and N a nonzero integer. There are five good methods for solving the general Pell's equation:

1. The Lagrange–Matthews–Mollin (LMM) method;
2. Brute-force search (which is good only if $|N|$ is small and the minimal positive solution to Pell's resolvent is small);
3. Use of quadratic rings;
4. The cyclic method;
5. Lagrange's system of reductions.

Of these five, we will present only the first three, with two versions for the third one. These two last algorithms are comparable in terms of effectiveness. For the cyclic method see [67] and for the Lagrange's system of reductions see [52] or [142].

4.1 General Theory

In a memoir of 1768, Lagrange gave a recursive method for solving the equation

$$x^2 - Dy^2 = N \qquad (4.1.1)$$

with $\gcd(x, y) = 1$, where $D > 1$ is not a perfect square and $N \neq 0$, thereby reducing the problem to the situation where $|N| < \sqrt{D}$, in which case the positive solutions (x, y) are found among the pairs (p_n, q_n), with p_n/q_n a convergent of the simple continued fraction for $\sqrt{D}$.

It does not seem to be widely known that Lagrange also gave another algorithm in a memoir of 1770, which may be regarded as a generalization of the well-known

© Springer Science+Business Media New York 2015
T. Andreescu, D. Andrica, *Quadratic Diophantine Equations*,
Developments in Mathematics 40, DOI 10.1007/978-0-387-54109-9_4

method of solving Pell's equation and negative Pell's equation presented in Section 3.3 by using the simple continued fraction for $\sqrt{D}$ (see [196]). Quadratic Reciprocity (see [21, 185] and [179] for various applications)

In what follows we will call (4.1.1) the *general Pell's equation*. First we will present the general method of finding the solutions to this equation following the presentation in [56, 112, 125, 126, 151] and [161].

Like in Section 3.5 we will consider the Pell's resolvent

$$u^2 - Dv^2 = 1. \tag{4.1.2}$$

Let $(u_n, v_n)_{n \geq 0}$ be the general solution to the equation (4.1.2) given in Theorem 3.2.1. Assume that equation (4.1.1) is solvable and let (x, y) be one of its solutions. Then

$$(u_n + v_n\sqrt{D})(x + y\sqrt{D}) = (u_nx + v_nyD) + (u_ny + v_nx)\sqrt{D}$$

and

$$(u_nx + v_nyD)^2 - D(u_ny + v_nx)^2 = (x^2 - Dy^2)(u_n^2 - Dv_n^2) = N \cdot 1 = N.$$

It follows that $(x_n, y_n)_{n \geq 0}$, where

$$x_n = xu_n + Dyv_n \quad \text{and} \quad y_n = yu_n + xv_n \tag{4.1.3}$$

satisfies the general Pell's equation. Hence every initial solution to (4.1.1) generates its own family of infinitely many solutions.

This method of generating solutions is called the *multiplication principle*.

The main problem here is to decide whether or not two different initial solutions generate different general solutions described above.

We say that solution $(x_n, y_n)_{n \geq 0}$ given by (4.1.3) is *associated* with the solution $(u_n, v_n)_{n \geq 0}$. The set of all solutions associated with each other forms a *class of solutions* to (4.1.1).

Next we will show a way to decide whether the two given solutions (x, y) and (x', y') belong to the same class or not. In fact, by using the method given in Theorem 3.5.2 it is easy to see that the necessary and sufficient condition for these two solutions to be associated with each other is that the numbers

$$\frac{xx' - Dyy'}{N} \quad \text{and} \quad \frac{yx' - xy'}{N}$$

are both integers.

Let K be the class consisting of the solutions $(x_n, y_n)_{n \geq 0}$ defined by (4.1.3). Then $(x_n, -y_n)_{n \geq 0}$ also constitutes a class, denoted by $\overline{K}$. The classes K and $\overline{K}$ are said to be *conjugates* of each other. Conjugate classes are in general distinct, but may sometimes coincide; in the latter case we speak of *ambiguous* classes.

Among all the solutions (x, y) in a given class K we now choose a solution (x^*, y^*) in the following way: let y^* be the least nonnegative value of y which occurs in K. If K is not ambiguous, then the number x^* is also uniquely determined, for the solution $(-x^*, y^*)$ belongs to the conjugate class $\overline{K}$. If K is ambiguous, then we get a uniquely determined x^* by prescribing that $x^* \geq 0$. The solution (x^*, y^*) defined in this way is said to be the *fundamental solution of the class*.

In the fundamental solution, the number $|x^*|$ has the least value which is possible for $|x|$ when (x, y) belongs to K. The case $x^* = 0$ can occur when the class is ambiguous, and similarly for the case $y^* = 0$.

If $N = \pm 1$, there is only one class and it is ambiguous.

Suppose now that N is positive.

Theorem 4.1.1. *If (x, y) is the fundamental solution of the class K of the equation (4.1.1) and if (u_1, v_1) is the fundamental solution of the Pell's resolvent (4.1.2), then the following inequalities hold:*

$$0 \leq |x| \leq \sqrt{\frac{(u_1 + 1)N}{2}} \tag{4.1.4}$$

$$0 < y \leq \frac{v_1}{\sqrt{2(u_1 + 1)}} \sqrt{N}. \tag{4.1.5}$$

Proof. If inequalities (4.1.4) and (4.1.5) are true for a class K, they are also true for the conjugate class $\overline{K}$. Thus we may assume that y is positive.

It is clear that

$$xu_1 - Dyv_1 = xu_1 - \sqrt{(x^2 - N)(u_1^2 - 1)} > 0. \tag{4.1.6}$$

Consider the solution $(xu_1 - Dyv_1, yu_1 - xv_1)$ which belongs to the same class as (x, y). Since (x, y) is the fundamental solution of the class and since by (4.1.6) $xu_1 - Dyv_1$ is positive, we must have $xu_1 - Dyv_1 \geq x$. From this inequality it follows that

$$x^2(u_1 - 1)^2 \geq D^2 y^2 v_1^2 = (x^2 - N)(u_1^2 - 1)$$

or

$$\frac{u_1 - 1}{u_1 + 1} \geq 1 - \frac{N}{x^2}$$

and finally $x^2 \leq \frac{1}{2}(u_1 + 1)N$. This proves inequality (4.1.4) and it is easily seen that (4.1.4) implies (4.1.5). □

Suppose next that $N < 0$ and call (4.1.1) the *general negative Pell's equation*. With a proof similar to the one in Theorem 4.1.1 we have

Theorem 4.1.2. *If (x, y) is the fundamental solution of the class K of the general negative Pell's equation and if (u_1, v_1) is the fundamental solution of the Pell's resolvent (4.1.2), then the following inequalities hold:*

$$0 \leq |x| \leq \sqrt{\frac{(u_1 - 1)|N|}{2}} \tag{4.1.7}$$

$$0 < y \leq \frac{v_1}{\sqrt{2(u_1 - 1)}} \sqrt{|N|}. \tag{4.1.8}$$

From Theorems 4.1.1 and 4.1.2 we deduce

Theorem 4.1.3. *If D is a nonsquare positive integer and N is a nonzero integer, then the equation (4.1.1) has a finite number of classes of solutions. The fundamental solutions of all the classes can be found after a finite number of trials by means of the inequalities (4.1.4), (4.1.5) and (4.1.7), (4.1.8). If (x^*, y^*) is the fundamental solution of the class K, then all the solutions in K are given by $(x_n, y_n)_{n \geq 0}$, where*

$$x_n = x^* u_n + D y^* v_n \quad \text{and} \quad y_n = y^* u_n + x^* v_n$$

and $(u_n, v_n)_{n \geq 0}$ represents the general solution of Pell's resolvent including ± 1, if necessary.

Remark. The upper bounds for fundamental solutions that generate the classes of solutions of general Pell's equation (4.1.1) found in Theorems 4.1.1 and 4.1.2 can still be improved. In [76] it is shown that

$$0 \leq |x| \leq \sqrt{\varepsilon |N|}, \quad 0 < y \leq \sqrt{\frac{\varepsilon |N|}{D}}$$

where $\varepsilon = u_1 + v_1 \sqrt{D}$.

In the private communication (L. Panaitopol, personal communication, December 2001) the following better upper bounds are mentioned

$$0 \leq |x| \leq \sqrt{\frac{|N| u_1 + N}{2}}, \quad 0 < y \leq \sqrt{\frac{|N| u_1 - N}{2D}}.$$

In the above delimitations (u_1, v_1) denotes the fundamental solution to the Pell's equation (4.1.2).

We denote by $k(D, N)$ the number of classes of solutions of the equation (4.1.1), and by $\mathcal{K}(D, N)$ the set of the fundamental solutions of all classes.

Theorem 4.1.4. *Let p be a prime. Then each of the general Pell's equations*

$$x^2 - D y^2 = \pm p \tag{4.1.9}$$

has at most one solution (x, y) in which x and y satisfy the inequalities (4.1.4) and (4.1.5), or (4.1.7) and (4.1.8), respectively, provided that $x \geq 0$.

If the equation (4.1.9) is solvable, then it has one or two solutions satisfying the above conditions, according as the prime p divides $2D$ or not.

Proof. Suppose that (x, y) and (x_1, y_1) are two solutions of (4.1.9) satisfying the conditions in the first part of Theorem 4.1.4. Thus the numbers x, y, x_1 and y_1 are nonnegative.

Eliminating D between relations

$$x^2 - Dy^2 = \pm p, \quad x_1^2 - Dy_1^2 = \pm p \qquad (4.1.10)$$

yields $x^2 y_1^2 - x_1^2 y^2 = \pm p(y_1^2 - y^2)$. Thus $xy_1 \equiv x_1 y \pmod{p}$.

Furthermore, from (4.1.10) we obtain

$$(xx_1 \mp Dyy_1)^2 - D(xy_1 \mp x_1 y)^2 = p^2.$$

In the equation

$$\left(\frac{xx_1 \mp Dyy_1}{p}\right)^2 - D\left(\frac{xy_1 \mp x_1 y}{p}\right)^2 = 1 \qquad (4.1.11)$$

let us choose the sign such that the congruence $xy_1 \equiv \pm x_1 y \pmod{p}$ is satisfied. Then the two squares on the left-hand side are integers. If $xy_1 \mp x_1 y \neq 0$, from (4.1.11) we conclude that

$$|xy_1 \mp x_1 y| \geq v_1 p. \qquad (4.1.12)$$

On the other hand, by applying inequalities (4.1.4) and (4.1.5), or (4.1.7) and (4.1.8), respectively, we obtain $|xy_1 \mp x_1 y| < v_1 p$, which is contrary to (4.1.12). The remaining case is $xy_1 \mp x_1 y = 0$, which is obviously possible only for $x = x_1$ and $y = y_1$. Thus the first part of Theorem 4.1.4.

Consequently, there are at most two classes of solutions. Suppose that (x, y) and $(x, -y)$ are two solutions which satisfy inequalities (4.1.4) and (4.1.5), or (4.1.7) and (4.1.8), respectively. These solutions are associated if and only if p divides the two numbers $2xy$ and $x^2 + Dy^2 = 2Dy^2 \pm p$. Since y cannot be divisible by p, the numbers $2x$ and $2D$ are divisible by p. But if $2D$ is divisible by p, then so is $2x$. Thus, the necessary and sufficient condition for (x, y) and $(x, -y)$ to belong to the same class is that $2D$ is a multiple of p. Thus proves the second part of the theorem. $\square$

The following example illustrates how the method described in Theorem 4.1.4 can be applied.

Consider the equation $x^2 - 2y^2 = 119$. The fundamental solution of its Pell's resolvent $u^2 - 2v^2 = 1$ is $(3, 2)$. The following solutions of our equation satisfy inequalities (4.1.4) and (4.1.5): $(11, 1), (-11, 1), (13, 5), (-13, 5)$. It is not difficult to show that these numbers are all fundamental solutions in different classes. Thus the number of classes is four but only solutions (11,1) and (13,5) satisfy Theorem 4.1.4. The form of the integer N is very important. For instance, in the paper [174] are considered the equations $x^2 - Dy^2 = \pm c(2^{31} - 1)$.

We will now present an example which illustrates how one can use Theorem 4.1.4. We will now rely on the result in our paper [16]. In [37] the following question is posed: Does the Diophantine equation

$$8x^2 - y^2 = 7 \tag{4.1.13}$$

have infinitely many solutions in positive integers?

Recently, in the paper [114] the more general equation $ax^2 - by^2 = c$ is considered. It is shown that if ab is not a square and the above equation has a positive integer solution (x_0, y_0), then it has infinitely many positive integer solutions. This property is a direct consequence of the multiplication principle. In the paper [143] a simple criterion for solving both equations $x^2 - Dy^2 = c$ and $x^2 - Dy^2 = -c$ is presented.

In what follows, we will find all solutions to the equation (4.1.13). We can write the equation (4.1.13) in the following equivalent form: $y^2 - 8x^2 = -7$. This is a special case of (4.1.9). In our case, $p = 7$ and p does not divide $2D = 16$. Applying Theorem 4.1.4 we deduce that the equation (4.1.13) has two classes of solutions and these are generated by $(-1, 1)$ and $(1, 1)$. The Pell's resolvent $u^2 - 8v^2 = 1$ has the fundamental solution $(u_1, v_1) = (3, 1)$ and its general solution $(u_n, v_n)_{n \geq 0}$ is given by (see formulas (3.2.6)):

$$\begin{cases} u_n = \dfrac{1}{2} \left[\left(3 + \sqrt{8}\right)^n + \left(3 - \sqrt{8}\right)^n \right] \\[4mm] v_n = \dfrac{1}{2\sqrt{8}} \left[\left(3 + \sqrt{8}\right)^n - \left(3 - \sqrt{8}\right)^n \right] \end{cases} \tag{4.1.14}$$

Applying Theorem 4.1.3 it follows that all solutions to the equation (4.1.13) are given by $(x_n, y_n)_{n \geq 0}$ and $(x'_n, y'_n)_{n \geq 0}$, where

$$\begin{cases} x_n = u_n + v_n \\ y_n = u_n + 8v_n \end{cases} \quad \text{and} \quad \begin{cases} x'_n = u_n - v_n \\ y'_n = -u_n + 8v_n \end{cases}$$

and $(u_n, v_n)_{n \geq 0}$ is defined in (4.1.14). We obtain two classes of solutions:

$$(x, y) = (1, 1), \ (4, 11), \ (23, 65), \ (134, 379), \ldots$$

and

$$(x', y') = (2, 5), \ (11, 31), \ (64, 181), \ (373, 1055), \ldots$$

respectively.

Remark. We will describe now the set of rational solutions to the Pell's equation $u^2 - Dv^2 = 1$. A family of such solutions was given in Remark 4 in Section 3.2.

For fixed positive integers m and n consider the general Pell's equation $x^2 - Dy^2 = (mn)^2$. Consider the set of all its integral solutions (x, y) satisfying $n|x$ and $m|y$ and let $S_{m,n}$ be the set of all pairs $\left(\dfrac{x_1}{m}, \dfrac{y_1}{n}\right)$, where $x = x_1 n$, $y = y_1 m$. The set of all rational solutions to $u^2 - Dv^2 = 1$ is then given by $S = \bigcup\limits_{m,n \geq 1} S_{m,n}$.

The following interesting result was proved in the paper [72].

Theorem 4.1.5. *Let $D = a^2 + (2b)^2$, with $a, b \in \mathbb{Z}$. If D is a prime, the following hold:*

1) The equation $x^2 - Dy^2 = a$ is solvable.
2) The equation $x^2 - Dy^2 = 4b$ is solvable.

If D is not prime, then both 1) and 2) can fail. For instance $221 = 10^2 + 11^2 = 5^2 + 14^2$, but for $a = \pm 5$ or ± 11 the equation $x^2 - 221y^2 = a$ has no solution mod 13, while for $b = \pm 5$ or ± 7 the equation $x^2 - 221y^2 = 4b$ has no solution mod 17.

4.2   Solvability of General Pell's Equation

Disregarding any time considerations, Theorem 4.1.1 may be used to determine whether any general Pell equation is solvable or not. Following the reference [204], let consider the general Pell equation $x^2 - 43y^2 = 35$. According to Theorem 4.1.1, if it is solvable, then any of its fundamental solutions (x, y) must lie within the following bounds: $0 < |x| \leq \left\lceil \sqrt{35(3482+1)/2} \right\rceil = 246$ and $0 < v \leq \left\lceil 532\sqrt{35/(2(3482+1))} \right\rceil = 37$, where $(3482, 532)$ is the fundamental solution to the Pell's resolvent $u^2 - 43v^2 = 1$. After checking all 9102 possible combinations of (x, y) we see that the equation $x^2 - 43y^2 = 35$ is not solvable.

With regards to computational efficiency, the question of solvability for the considered example is no match for modern computers. But, what happens when N gets large? Clearly, $\sqrt{\dfrac{(u_1 - 1)|N|}{2}} \to \infty$, as $N \to \pm\infty$. Thus, Theorem 4.1.1, though a nice tool, does not allow one to efficiently decide if a particular general Pell equation is solvable. In the reference [204] is mentioned the equation $x^2 - 313y^2 = 172635965$ and the fact that, using the actual computation force, we need about 69806785 years to prove the unsolvability, following the method provided by Theorem 4.1.1. Thus, in this particular example, with $N = 172635965$ relatively small, using the approach in Theorem 4.1.1 will take a considerable amount of time.

The question of solvability of the general Pell's equation can be formulated into two problems:

Pell Decision Problem (PDP). *Given a positive integer $D \geq 2$ which is not a perfect square, and an integer N, is there an efficient means to decide if the equation (4.1.1) is solvable?*

In some situations PDP can be reduced to the case when D is a prime (see for instance Theorem 3.6.2).

Pell Search Problem (PSP). *Assuming that the equation (4.1.1) is solvable, can we find all fundamental solutions in the Pell classes in a reasonable amount of time?*

Notice that a general criterion for solvability is, in effect, a solution to a PDP. In this section we address the problem of finding a general criterion for solvability of general Pell equation and give a partial solution. Most of the tests that we develop throughout this section are based on the reference [204] and do not rely on integer factorization. However, a few of the implementations based on these results will rely heavily on the efficiency of integer factorization, which is likely no more efficient than tests based on the Pell class approach.

4.2.1 PDP and the Square Polynomial Problem

In what follows we will show that the PDP is equivalent to the problem of deciding whether or not a particular second degree polynomial with integer coefficients has a square integer value. In this respect we formulate the following concrete problem:

Square Polynomial Decision Problem (SPDP). *Does there exist an algorithm that, for any odd prime p and $N \in \mathbb{Z}$ with $\gcd(N, p) = 1$ decides if there is for some a with $N \equiv a^2 \pmod{p}$ and $n \in \mathbb{Z}$ such that $pn^2 - 2an + \dfrac{a^2 - N}{p}$ is a square?*

The SPDP and the PDP for specific D and N may be formulated in terms of arithmetical functions. Let p be a prime, $N \in \mathbb{Z}$, with $\left(\dfrac{N}{p}\right) = 1$ and consider the equation $x^2 - py^2 = N$. Since $\left(\dfrac{N}{p}\right) = 1$ we have $a^2 \equiv N \pmod{p}$ for some positive integer a. Note that, the Tonelli–Shanks algorithm, assuming the Generalized Riemann Hypothesis, efficiently find an integer a such that $a^2 \equiv N \pmod{p}$ (see [159, pp. 110–115]).

Following the reference [204], define the functions

$$\phi(p, N) = \begin{cases} 1 \text{ if } pn^2 - 2an + \dfrac{a^2 - N}{p} \text{ is a square for some} \\ \quad \text{integers } n \text{ and } a \text{ with } a^2 \equiv N \pmod{p} \\ -1 \text{ otherwise} \end{cases}$$

and

$$\psi(p,N) = \begin{cases} 1 \text{ if } x^2 - py^2 = N \text{ is solvable} \\ -1 \text{ otherwise.} \end{cases}$$

Clearly, in the definition of $\phi(p,N)$ we can assume that $a \in \mathbb{Z}_p$.

Now, we have the proper terminology to prove the following result.

Theorem 4.2.1. *Let p be an odd prime and $N \in \mathbb{Z}$ with $\left(\dfrac{N}{p}\right) = 1$. Then, $\psi(p,N) = 1$ if and only if $\phi(p,N) = 1$.*

Proof. If $\psi(p,N) = 1$, then we have $u^2 - pm^2 = N$ for some integers u, m. Observe that $y^2 \equiv N \pmod{p}$. To prove $\phi(p,N) = 1$ we must show that there is an integer n such that $pn^2 - 2un + \dfrac{u^2 - N}{p}$ is a square. But, we have $\dfrac{u^2 - N}{p} = m^2$, hence we choose $n = 0$. Therefore $\phi(p,N) = 1$.

Suppose $\phi(p,N) = 1$. That is, there are integers n, m and $u \in \mathbb{Z}_p$ such that $u^2 \equiv N \pmod{p}$ and $pn^2 - 2un + \dfrac{u^2 - N}{p} = m^2$. The last relation is equivalent to $(u + pn)^2 - pm^2 = N$, so the pair $(u + pn, m)$ is a solution to the general Pell's equation $x^2 - py^2 = N$. Thus, the relation $\psi(p,N) = 1$ holds. $\qquad\square$

The result in Theorem 4.2.1 proves that SPDP is equivalent to PDP.

4.2.2 The Legendre Test

The Legendre symbol and the Quadratic Reciprocity Law provide the first test for the solvability of general Pell's equation.

Theorem 4.2.2. *If $\left(\dfrac{N}{p}\right) = -1$ then $\psi(p,N) = -1$, that is $x^2 - py^2 = N$ is not solvable.*

Proof. If the equation $x^2 - py^2 = N$ were solvable, then $u^2 - pv^2 = N$ for some integers u and v. Therefore, $u^2 - N = pv^2$, hence $u^2 \equiv N \pmod{p}$, implying $\left(\dfrac{N}{p}\right) = 1$, contradicting our assumption. $\qquad\square$

Corollary 4.2.3. *If $\left(\dfrac{N}{p}\right) = 1$ and $\left(\dfrac{M}{p}\right) = -1$, then the equation*

$$x^2 - py^2 = MN$$

is not solvable.

Example 1. Consider the general Pell's equation $x^2 - 17y^2 = -46$. Using the properties of the Legendre symbol, we have

$$\left(\frac{-46}{17}\right) = \left(\frac{-1}{17}\right)\left(\frac{46}{17}\right) = \left(\frac{12}{17}\right) = \left(\frac{4}{17}\right)\left(\frac{3}{17}\right)$$

$$= \left(\frac{3}{17}\right) = \left(\frac{17}{3}\right) = \left(\frac{2}{3}\right) = -1,$$

hence, according to Theorem 4.2.2, the considered equation is not solvable.

4.2.3 Legendre Unsolvability Tests

This subsection uses the Quadratic Reciprocity Law and some properties of the Legendre symbol to obtain some tests for the unsolvability of general Pell's equation.

Theorem 4.2.4. *Let p be an odd prime and N a positive integer. If $p \equiv 3 \pmod 4$ and $\left(\dfrac{N}{p}\right) = 1$, then the equation $x^2 - py^2 = -N$ is not solvable.*

Proof. If the equation is solvable, then we have $r^2 - ps^2 = -N$, for some integers r, s. It follows $r^2 \equiv -N \pmod p$, hence $\left(\dfrac{-N}{p}\right) = 1$. Using the standard properties of the Legendre symbol we get

$$1 = \left(\frac{-N}{p}\right) = \left(\frac{-1}{p}\right)\left(\frac{N}{p}\right)$$

and

$$\left(\frac{-1}{p}\right) = \left(\frac{-1}{p}\right) \cdot 1 = \left(\frac{-1}{p}\right)\left(\frac{N}{p}\right) = \left(\frac{-N}{p}\right) = 1.$$

By the Quadratic Reciprocity Law, this only happens when $p \equiv 1 \pmod 4$, contradicting our assumption. □

Example 2. Consider the equation $x^2 - 11y^2 = -5$. Since $(4,1)$ is a solution to $x^2 - 11y^2 = 5$, we have $\left(\dfrac{5}{11}\right) = 1$. Since $11 \equiv 3 \pmod 4$, by Theorem 4.2.4, we obtain that the considered equation is not solvable.

The next result is given in [204] and it yields a general test for the unsolvability of a large class of general Pell's equations.

Theorem 4.2.5. *Let p be a prime, $p \equiv 3 \pmod 4$, and $N = m^2 n$ with n square free. If $x^2 - py^2 = N$ is solvable, then $n \equiv 1 \pmod 4$.*

Proof. Suppose that $n \equiv 3 \pmod 4$. Since $x^2 - py^2 = N$ is solvable, there are $u, v \in \mathbb{Z}$ such that $u^2 - pv^2 = m^2 n$. We shall collect the following three facts:

(i) $\left(\dfrac{n}{p}\right) = 1$.

(ii) If q is a prime divisor of n with $q \equiv 3 \pmod 4$, then $\left(\dfrac{q}{p}\right) = -1$.

(iii) Let $r = |\{q : q | n \text{ and } q \text{ is an odd prime and } q \equiv 3 \pmod 4\}|$. Then r is odd.

Since $u^2 - pv^2 = m^2 n$, we have $u^2 - m^2 n = pv^2$. So, $X^2 \equiv N \pmod p$ is solvable. Thus, $\left(\dfrac{n}{p}\right) = \left(\dfrac{m^2 n}{p}\right) = 1$. This proves (i).

Now suppose that q is a prime divisor of n. So, $n = qn_0$ for some $n_0 \in \mathbb{Z}$. Thus, $u^2 - pv^2 = qn_0$ and so $X^2 \equiv pv^2 \pmod p$ is solvable. So, $\left(\dfrac{p}{q}\right) = \left(\dfrac{pv^2}{q}\right) = 1$. By the Quadratic Reciprocity Law, we know that, since $p \equiv q \equiv 3 \pmod 4$, $\left(\dfrac{p}{q}\right) = -\left(\dfrac{q}{p}\right)$. So, $-\left(\dfrac{q}{p}\right) = 1$. This proves (ii).

Let $n = q_1 \dots q_r \cdot q_{r+1} \dots q_l$, where $q_1 \equiv \dots \equiv q_r \equiv 3 \pmod 4$ and $q_{r+1} \equiv \dots \equiv q_l \equiv 1 \pmod 4$. If r is even, then we may arrange these first r primes in pairs as follows: $(q_1 \cdot q_2), (q_3 \cdot q_4), \dots, (q_{r-1} \cdot q_r)$. Then, for i even with $1 \leq i \leq r$, $(q_{i-1} \cdot q_i) \equiv 9 \equiv 1 \pmod 4$. But, then $n = (q_1 \cdot q_2) \cdot (q_3 \cdot q_4) \dots (q_{r-1} \cdot q_r) \cdot q_{r+1} \dots q_l \equiv 1 \pmod 4$ contrary to assumption. This proves (iii).

Again let $n = q_1 \dots q_r q_{r+1} \dots q_l$, where $q_1 \equiv \dots \equiv q_r \equiv 3 \pmod 4$ and $q_{r+1} \equiv \dots \equiv q_l \equiv 1 \pmod 4$. Because r is odd, by (ii) we have

$$\left(\frac{q_1}{p}\right) \dots \left(\frac{q_r}{p}\right) = (-1) \dots (-1) = (-1)^r = -1.$$

Also,

$$\left(\frac{q_{r+1}}{p}\right) \dots \left(\frac{q_l}{p}\right) = 1 \dots 1 = 1^{l-r} = 1.$$

Therefore, by (i) we obtain

$$1 = \left(\frac{n}{p}\right) = \left(\frac{q_1 \dots q_r q_{r+1} \dots q_l}{p}\right)$$

$$= \left(\frac{q_1}{p}\right) \dots \left(\frac{q_r}{p}\right) \left(\frac{q_{r+1}}{p}\right) \dots \left(\frac{q_l}{p}\right) = (-1)^r \cdot 1 = -1,$$

a contradiction. $\square$

The result in Theorem 4.2.5, when expressed using the contrapositive, yields a nice test for unsolvability. We state this as the following consequence.

Corollary 4.2.6. *Let $N = m^2n$ with n square free. If p is a prime with $p \equiv n \equiv 3$ (mod 4), then the equation $x^2 - py^2 = N$ is not solvable.*

The next consequence follows immediately.

Corollary 4.2.7. *If $p \equiv N \equiv 3$ (mod 4) and $M \equiv 1$ (mod 4), then the equation $x^2 - py^2 = MN$ is not solvable.*

Example 3. Consider the equation $x^2 - 31y^2 = 1008$. We have $1008 = 12^2 \cdot 7$ and $7 \equiv 3$ (mod 4). Corollary 4.2.6 allows us to conclude that the equation is not solvable.

The next result requires that we know a prime factor ≥ 3 of N.

Theorem 4.2.8. *Let q be an odd prime divisor of N. If the equation $x^2 - py^2 = N$ is solvable, then $\left(\dfrac{p}{q}\right) = 1$.*

Proof. Assume that $u^2 - pv^2 = N$, for some integers u, v. Because $N \equiv 0$ (mod q), it follows $u^2 \equiv pv^2$ (mod q).

Therefore, $\left(\dfrac{pv^2}{q}\right) = 1$, hence $\left(\dfrac{p}{q}\right) = 1$.   □

In the case when we can find an odd prime divisor of N, the contrapositive to Theorem 4.2.8 provides a nice test for unsolvability.

Corollary 4.2.9. *Let q be an odd prime divisor of N. If $\left(\dfrac{p}{q}\right) = -1$, then the equation $x^2 - py^2 = N$ is not solvable.*

Now, we are in position to discuss the solvability of the equation

$$x^2 - 313y^2 = 172635965,$$

considered at the beginning of this section. Because 5 is a prime divisor of 172635965 and $\left(\dfrac{313}{5}\right) = -1$, we may use Corollary 4.2.9 to conclude the unsolvability of the equation.

Corollary 4.2.10. *Let q be an odd prime divisor of N. If p or $q \equiv 1$ (mod 4) and $\left(\dfrac{q}{p}\right) = -1$, then the equation $x^2 - py^2 = N$ is not solvable.*

Proof. Because p or $q \equiv 1$ (mod 4), we have $\left(\dfrac{p}{q}\right) = \left(\dfrac{q}{p}\right) = -1$, and we can use the result in Corollary 4.2.9.

Corollary 4.2.11. *Let $N = m^2n$, where n is square free. If p is a prime with $p \equiv 5$ (mod 8), and n is even, then the equation $x^2 - py^2 = N$ is not solvable.*

Proof. Suppose that $u^2 - pv^2 = N$, for some integers u, v. It follows $\left(\dfrac{N}{p}\right) = 1$, hence $\left(\dfrac{n}{p}\right) = \left(\dfrac{m^2 n}{p}\right) = 1$. Since n is even, we have $n = 2n_0$, $n_0 \in \mathbb{Z}$, and $\left(\dfrac{2n_0}{p}\right) = 1$. Because $p \equiv 5 \pmod{8}$, we have $\left(\dfrac{2}{p}\right) = -1$. Therefore, n has a prime factor q such that $\left(\dfrac{q}{p}\right) = -1$. But, n is square free, so q must be odd. Moreover, from $p \equiv 5 \pmod{8}$, it follows $p \equiv 1 \pmod{4}$ and the conclusion follows from Corollary 4.2.9. $\qquad\square$

The following application is given in the reference [204].

Example 4. Consider the equation $x^2 - 181y^2 = 1908360$. We have $181 \equiv 5 \pmod{8}$ and $1908360 = 18^2 \cdot 5890$ with $5890 = 2 \cdot 5 \cdot 19 \cdot 31$ even and square free. Applying Corollary 4.2.11, it follows the unsolvability of the equation.

4.2.4 Modulo n Unsolvability Tests

We will describe a simple but useful way to test the unsolvability of general Pell's equation.

Theorem 4.2.12. *If the equation $x^2 - Dy^2 = N$ is not solvable in $\mathbb{Z}_n$, for some positive integer $n \geq 2$, then it is not solvable in integers.*

Proof. Assume that $u^2 - Dv^2 = N$ for some integers u, v. The remainder upon dividing $u^2 - Dv^2$ by n will be the same as the remainder in the division of N by n. Therefore, the equation $x^2 - Dy^2 = N$ is solvable in $\mathbb{Z}_n$. Thus, by the contrapositive, the result follows. $\qquad\square$

Note that if the equation $x^2 - Dy^2 = N$ is solvable in $\mathbb{Z}_n$ for some positive integer $n \geq 2$, then it is not necessarily the case that it is solvable in integers.

Theorem 4.2.13. *Let p be a prime with $p \equiv 3 \pmod{4}$, and N an odd integer. If the equation $x^2 - py^2 = N$ is solvable, then $N \equiv 1 \pmod{4}$.*

Proof. We have $a^2 \equiv 0\,1 \pmod{4}$, and by direct calculation, we see that $x^2 - oy^2 \equiv x^2 - 3y^2 \equiv 0, 1,$ or $2 \pmod{4}$. Therefore, if $N \equiv 3 \pmod{4}$, the equation is not solvable in $\mathbb{Z}_4$, and, by mod 4 test, the equation is not solvable in integers. $\qquad\square$

Using the same argument, but in $\mathbb{Z}_8$, we can prove the following result.

Theorem 4.2.14. *Let p be a prime with $p \equiv 1, 3,$ or $5 \pmod{8}$, and N an integer with $N \equiv 2 \pmod{4}$. Then the equation $x^2 - py^2 = N$ is not solvable.*

4.2.5 Extended Multiplication Principle

We now give some tests for the solvability of general Pell's equation, using an extension of the multiplication principle discussed in Section 4.1.

Extended Multiplication Principle. *If* $u^2 - Dv^2 = M$ *and* $r^2 - Ds^2 = N$, *then* $(ur \pm Dvs)^2 - D(us \pm vr)^2 = MN$, *where the signs* $+$ *and* $-$ *correspond.*

The above identity is also called the *Bhaskara identity*, according to the name of the Hindu mathematician mentioned in Section 3.1.

We can reformulate this algebraic property as follows: If the general Pell's equations $u^2 - Dv^2 = M$ and $r^2 - Ds^2 = N$ are solvable, then the equation $x^2 - Dy^2 = MN$ is also solvable.

As an application to the Extended Multiplication Principle, we present an extension of the result involving the negative Pell's equation, and contained in Theorem 3.6.2.

Theorem 4.2.15. *Let* p *be a prime with* $p \equiv 1 \pmod 4$. *The equation*

$$x^2 - py^2 = -N$$

is solvable if and only if the equation $x^2 - py^2 = N$ *is solvable.*

Proof. By Theorem 3.6.2, we know that the negative Pell's equation

$$u^2 - pv^2 = -1$$

is solvable. Now, the result directly follows from the Extended Multiplicative Principle. □

Remark. Notice that if we can factor N, say $N = N_1 \ldots N_s$ and show, for all $i = 1, \ldots, s$, that the equation $x^2 - Dy^2 = N_i$ is solvable, then using successively the Extended Multiplication Principle we obtain that the equation $x^2 - Dy^2 = N$ is solvable.

The converse of the above remark need not hold, as the next example illustrates.

Example 5. Considering the equation $x^2 - 37y^2 = 192$, we have $192 = 4^2 \cdot 12 = 8^2 \cdot 3$. The equations $x^2 - 37y^2 = 12$ and $x^2 - 37y^2 = 4^2$ are clearly solvable, hence according to the Extended Multiplication Principle, it follows the considered equation is solvable. On the other hand, if we use the second factorization of 192, we see that $x^2 - 37y^2 = 3$ is not solvable (apply Theorem 4.1.1 where $(u_1, v_1) = (73, 12)$. Thus, we may not conclude that the unsolvability of $x^2 - 37y^2 = 3$ implies the unsolvability of $x^2 - 37y^2 = 192$.

4.3   An Algorithm for Determining the Fundamental Solutions Based on Simple Continued Fractions (The LMM Method)

We will describe an almost forgotten algorithm due to Lagrange, for deciding the solvability of general Pell's equation (4.1.1), where $\gcd(x, y) = 1$ and $D > 0$ is not a perfect square. In the case of solvability, the fundamental solutions are also constructed.

The main purpose of this section is to present a version of Lagrange's algorithm which uses only the technique of simple continued fractions.

A related algorithm is given in [158] but each of the cases $D = 2$ or $D = 3$ and $N < 0$ needs separate consideration. Also, unlike our algorithm, the approach in [158] requires the calculation of the fundamental solution of Pell's resolvent.

Lagrange's algorithm has been rediscovered in [141]. The method there is more complicated than ours, as it uses the language of ideals and semi-simple continued fractions, in addition to that of simple continued fractions.

First we need a result which is an extension of Theorem 172 in [88].

Lemma 4.3.1. *If* $\omega = \dfrac{P\zeta + R}{Q\zeta + S}$, *where* $\zeta > 1$ *and* P, Q, R, S *are integers such that* $Q > 0$, $S > 0$ *and* $PS - QR = \pm 1$, *or* $S = 0$ *and* $Q = R = 1$, *then* P/Q *is a convergent to* ω. *Moreover if* $Q \neq S > 0$, *then*

$$\frac{R}{S} = \frac{p_{n-1} + kp_n}{q_{n-1} + kq_n}, \quad k \geq 0.$$

Also, $\zeta + k$ *is the* $(n + 1)$*-th complete convergent to* ω. *Here* $k = 0$ *if* $Q > S$, *while* $k \geq 1$ *if* $Q < S$.

Proof. In [88] only the case $Q > S > 0$ is considered. We write

$$\frac{P}{Q} = \langle a_0; a_1, \ldots, a_n \rangle = \frac{p_n}{q_n}$$

and assume $PS - QR = (-1)^{n-1}$. Then

$$p_n S - q_n R = PS - QR = p_n q_{n-1} - p_{n-1} q_n,$$

so $p_n(S - q_{n-1}) = q_n(R - p_{n-1})$.

Hence $q_n | (S - q_{n-1})$. Then from $q_n = Q > S > 0$ and $q_n \geq q_{n-1} > 0$, we deduce $|S - q_{n-1}| < q_n$ and hence $S - q_{n-1} = 0$. Then $S = q_{n-1}$ and $R = p_{n-1}$.

Also

$$\omega = \frac{P\zeta + R}{Q\zeta + S} = \frac{p_n \zeta + p_{n-1}}{q_n \zeta + q_{n-1}} = \langle a_0; a_1, \ldots, a_n, \zeta \rangle.$$

If $S = 0$ and $Q = R = 1$, then $\omega = [P, \zeta]$ and $P/Q = P/1 = p_0/q_0$.

If $Q = S$, then $Q = S = 1$ and $P - R = \pm 1$. If $P = R + 1$, then $\omega = [R, 1, \zeta]$, so $P/Q = (R + 1)/1 = p_1/q_1$. If $P = R - 1$, then $\omega = [R - 1, 1 + \zeta]$ and $P/Q = (R - 1)/1 = p_0/q_0$.

If $Q < S$, then from $q_n | (S - q_{n-1})$ and

$$S - q_{n-1} > Q - q_{n-1} = q_n - q_{n-1} \geq 0,$$

we have $S - q_{n-1} = kq_n$, where $k \geq 1$. Then

$$\omega = \frac{P\zeta + R}{Q\zeta + S} = \frac{p_n\zeta + p_{n-1} + kp_n}{q_n\zeta + q_{n-1} + kq_n} = \frac{p_n(\zeta + k) + p_{n-1}}{q_n(\zeta + k) + q_{n-1}}$$

and $\omega = \langle a_0; a_1, \ldots, a_n, \zeta + k \rangle$.   □

Theorem 4.3.2. *Suppose* $x^2 - Dy^2 = N$ *is solvable in integers* $x > 0$, $y > 0$, *with* $\gcd(x, y) = 1$ *and let* $Q_0 = |N|$. *Then* $\gcd(Q_0, y) = 1$. *Define* P_0 *by* $x \equiv -P_0 y$ $(\mathrm{mod}\ Q_0)$, *where* $D \equiv P_0^2$ $(\mathrm{mod}\ Q_0)$ *and* $-Q_0/2 < P_0 \leq Q_0/2$.

Let $\omega = (P_0 + \sqrt{D})/Q_0$ *and let* $x = Q_0 X - P_0 y$. *Then*

(i) X/y *is a convergent* A_{n-1}/B_{n-1} *of* ω *if* $x > 0$;
(ii) $Q_n = (-1)^n N/|N|$.

Proof. With $Q_0 = |N|$, $x = Q_0 X - P_0 y$ and $x^2 - Dy^2 = N$, we have

$$P_0 x + Dy \equiv -P_0^2 y + Dy \equiv (-P_0^2 + D)y \equiv 0 \pmod{Q_0}.$$

Hence the matrix

$$\begin{bmatrix} P & R \\ Q & S \end{bmatrix} = \begin{bmatrix} X & \frac{P_0 x + Dy}{Q_0} \\ y & x \end{bmatrix}$$

has integer entries and determinant $\Delta = \pm 1$. For

$$\Delta = Xx - \frac{y(P_0 x + Dy)}{Q_0}$$

$$= \frac{(x + P_0 y)x}{Q_0} - \frac{y(P_0 x + Dy)}{Q_0}$$

$$= \frac{x^2 - Dy^2}{Q_0} = \pm 1.$$

Also, if $\zeta = \sqrt{D}$ and $\omega = (P_0 + \sqrt{D})/Q_0$, it is easy to verify that $\omega = \frac{P\zeta + R}{Q\zeta + S}$. Then the lemma implies that X/y is a convergent to ω.

Finally, $x = Q_0 X - P_0 y = Q_0 A_{n-1} - P_0 B_{n-1} = G_{n-1}$ and

$$N = x^2 - Dy^2 = G_{n-1}^2 - DB_{n-1}^2 = (-1)^n Q_0 Q_n.$$

Hence $Q_n = (-1)^n N/|N|$.   □

Remark. The solutions u of $u^2 \equiv D \pmod{Q_0}$ come in pairs $\pm u_1, \ldots, \pm u_r$, where $0 < u_i \leq Q_0/2$, together with possibly $u_{r+1} = 0$ and $u_{r+2} = Q_0/2$. Hence we can state the following:

Corollary 4.3.3. *Suppose $x^2 - Dy^2 = N$ is solvable, with $x > 0$ and $y > 0$, $\gcd(x, y) = 1$ and $Q_0 = |N|$. Let $x \equiv -P_0 y \pmod{Q_0}$, where $P_0 \equiv \pm u_i \pmod{Q_0}$ and $x = Q_0 X - P_0 y$. Then X/y is a convergent A_{n-1}/B_{n-1} of $\omega_i = (u_i + \sqrt{D})/Q_0$ or $\omega_i' = (-u_i + \sqrt{D})/Q_0$ and $Q_n = (-1)^n N/|N|$.*

4.3.1 An Algorithm for Solving the General Pell's Equation (4.1.1)

In view of the Corollary 4.3.3 we know that the primitive solutions to $x^2 - Dy^2 = N$ with $y > 0$ will be found by considering the continued fraction expansions of both ω_i and ω_i' for $1 \leq i \leq r + 2$.

One can show that each equivalence class contains solutions (x, y) with $x > 0$ and $y > 0$, so the necessary condition $Q_n = (-1)^n N/|N|$ occurs in both ω_i and ω_i'. Hence we need only consider ω_i.

Suppose that $\omega_i = (u_i + \sqrt{D})/Q_0 = [a_0, \ldots, a_t, \overline{a_{t+1}, \ldots, a_{t+l}}]$.

If $x^2 - Dy^2 = N$ is solvable, there are infinitely many solutions and hence $Q_n = \pm 1$ holds for ω_i for some n in the range $t + 1 \leq n \leq t + l$. Any such n must have $Q_n = 1$, as $(P_n + \sqrt{D})/Q_n$ is reduced for n in this range and so $Q_n > 0$. Moreover, if l is even, then the condition $(-1)^n = N/|N|$ is preserved.

In addition, there can be at most one such n. For if $P_n = \sqrt{D}$ is reduced, then $P_n = [\sqrt{D}]$ and hence two such occurrences of $Q_n = 1$ within a period would give a smaller period.

We also remark that l is odd if and only if the fundamental solution of Pell's equation has norm equal to -1. Consequently, a solution of $x^2 - Dy^2 = N$ gives rise to a solution of $x^2 - Dy^2 = -N$; indeed we see that if $t + 1 \leq n \leq t + l$ and $k \geq 1$, then $G_{n+kl-1} + B_{n+kl-1}\sqrt{D} = \eta_0^k(G_{n-1} + B_{n-1}\sqrt{D})$, where η_0 is the fundamental solution of $x^2 - Dy^2 = \pm 1$. Hence $G_{n+l-1}^2 - DB_{n+l-1}^2 = -(G_{n-1}^2 - DB_{n-1}^2)$ if $N(\eta_0) = -1$.

Putting these observations together, we have the following:

Theorem 4.3.4. *For $1 \leq i \leq r + 2$, let*

$$\omega_i = (u_i + \sqrt{D})/Q_0 = \langle a_0, \ldots, a_t, \overline{a_{t+1}, \ldots, a_{t+l}} \rangle.$$

(a) *Then a necessary condition for $x^2 - Dy^2 = N$, $\gcd(x,y) = 1$, to be solvable is that for some i in $i = 1, \ldots, r+2$, we have $Q_n = 1$ for some n in $t+1 \le n \le t+l$, where if l is even, then $(-1)^n N/|N| = 1$.*

(b) *Conversely, suppose for ω_i, we have $Q_n = 1$ for some n with $t + 1 \le n \le t+l$. Then*

    (i) *If l is even and $(-1)^n N/|N| = 1$, then $x^2 - Dy^2 = N$ is solvable and it has solution $G_{n-1} + B_{n-1}\sqrt{D}$.*

    (ii) *If l is odd, then $G_{n-1} + B_{n-1}\sqrt{D}$ is a solution of $x^2 - Dy^2 = (-1)^n|N|$, while $G_{n+l-1} + B_{n+l-1}\sqrt{D}$ is a solution of $x^2 - Dy^2 = (-1)^{n+1}|N|$.*

    (iii) *At least one of the $G_{n-1} + B_{n-1}\sqrt{D}$ with least B_{n-1} satisfying $Q_n = (-1)^n N/|N|$, which arise from continued fraction expansions of ω_i and ω_i', is a fundamental solution.*

Remarks. 1) Unlike the case of Pell's equation, $Q_n = \pm 1$ can also occur for $n < t + 1$ and can contribute to a fundamental solution. If $N(\eta) = 1$, one sees that to find the fundamental solutions for both $x^2 - Dy^2 = \pm N$, it suffices to examine only the cases $Q_n = \pm 1$, $n \le t + l$. However if $N(\eta) = -1$, one may have to examine the range $t + l + 1 \le n \le t + 2l$ as well.

2) It can happen that l is even and that $x^2 - Dy^2 = N$ is solvable and has solution $x \equiv \pm u_i y \pmod{Q_0}$, while $x^2 - Dy^2 = -N$ is solvable and has solution $x \equiv \pm u_j y \pmod{Q_0}$, with $i \ne j$. (Of course, if $|N| = p$ is prime, this cannot happen, as the congruence $u^2 \equiv D \pmod{p}$ has two solutions if p does not divide D and one solution if p divides D.)

An example of this is $D = 221$, $N = 217$ (see Example 2 later). Then $u_1 = 2$, $u_2 = 33$. Also, $l = 6$ and $(2+\sqrt{221})/217$ produces the solution $-2+\sqrt{221}$ of $x^2 - 221y^2 = -217$, whereas $(33 - \sqrt{221})/217$ produces the solution $-179 + 12\sqrt{221}$ of $x^2 - 221y^2 = 217$.

Example 1 (Lagrange). $x^2 - 13y^2 = \pm 101$.

We find the solutions of $P_0^2 \equiv 13 \pmod{101}$ are ± 35.

(a) We have $\dfrac{35 + \sqrt{13}}{101} = [0, 2, 1, \overline{1, 1, 1, 1, 6}]$.

i	0	1	2	3	4	5	6	7	8
P_i	35	-35	11	-2	3	1	2	1	3
Q_i	101	-12	9	1	4	3	3	4	1
A_i	0	1	1	2	3	5	8	13	86
B_i	1	2	3	5	8	13	21	34	225

We observe that $Q_3 = Q_8 = 1$. The period length is odd, so both the equations $x^2 - 13y^2 = \pm 101$ are solvable. With $G_n = Q_0 A_n - P_0 B_n$, we have

$$G_2 = 101 \cdot 1 - 34 \cdot 3 = -4, \quad x+y\sqrt{13} = -4+3\sqrt{13}, \quad x^2 - 13y^2 = -101;$$

$$G_7 = 101 \cdot 13 - 35 \cdot 34 = 123, \quad x+y\sqrt{13} = 123+34\sqrt{13}, \quad x^2 - 13y^2 = 101.$$

(b) We have $\dfrac{-35 + \sqrt{13}}{101} = [-1, 1, 1, 2, 4, \overline{1, 1, 1, 1, 6}]$.

i	0	1	2	3	4	5	6	7	8
P_i	−35	−66	23	1	3	1	2	1	3
Q_i	101	−43	12	1	4	3	3	4	1
A_i	−1	0	−1	−4	−5	−9	−14	−23	−152
B_i	1	1	3	13	16	29	45	74	489

We observe that $Q_3 = Q_8 = 1$. Hence

$$G_2 = 101 \cdot (-1) - (-35) \cdot 3 = 4, \quad x + y\sqrt{13} = A + 3\sqrt{13}, \quad x^2 - 13y^2 = -101;$$

$$G_7 = 101 \cdot (-23) - (-35) \cdot 74 = 267, \quad x + y\sqrt{13} = 267 + 74\sqrt{13}, \quad x^2 - 13y^2 = 101.$$

Hence $-4 + 3\sqrt{13}$ and $123 + 34\sqrt{13}$ are fundamental solutions for the equations $x^2 - 13y^2 = -101$ and $x^2 - 13y^2 = 101$ respectively.

We have $\eta = 649 + 180\sqrt{13}$, so the complete solution of $x^2 - 13y^2 = -101$ is given by $x + y\sqrt{13} = \pm\eta^n(\pm 4 + 3\sqrt{13})$, $n \in \mathbb{Z}$, while the complete solution of $x^2 - 13y^2 = 101$ is given by $x + y\sqrt{13} = \pm\eta^n(\pm 123 + 34\sqrt{13})$, $n \in \mathbb{Z}$.

Example 2. $x^2 - 221y^2 = \pm 217$.

We find the solutions of $P_0^2 \equiv 221 \pmod{217}$ are ± 2 and ± 33.

(a) We have $\dfrac{2 + \sqrt{221}}{217} = [0, 12, \overline{1, 6, 2, 6, 1, 28}]$.

i	0	1	2	3	4	5	6	7
P_i	2	−2	14	11	13	13	11	14
Q_i	217	1	25	4	13	4	25	1
A_i	0	1	1	7	15	97	112	3233
B_i	1	12	13	90	193	1248	1441	41596

We observe that $Q_1 = Q_7 = 1$. The period length is even and $(-1)^7 = -1$. Hence the equation $x^2 - 221y^2 = -217$ is solvable.

$$G_0 = 217 \cdot 0 - 2 \cdot 1 = -2, \quad x + y\sqrt{221} = -2 + \sqrt{221}, \quad x^2 - 221y^2 = -217.$$

i	0	1	2	3	4	5	6	7	8
P_i	33	−33	13	5	7	8	7	3	4
Q_i	101	−10	9	6	5	3	10	7	9

We see that the condition $Q_n = 1$ does not holds for $3 \leq n \leq 8$.

4.4   Solving the General Pell's Equation

4.4.1   The PQa Algorithm for Solving Pell's and Negative Pell's Equations

This algorithm is at the heart of all the algorithms to solve Pell's equations presented here. The input to the algorithms is three integers, D, P_0, Q_0, where $D > 0$ is not a square, $Q_0 > 0$, and $P_0^2 \equiv D \pmod{Q_0}$. Recursively compute, for $i \geq 0$

$$a_i = \text{int}(P_i + \sqrt{D})/Q_i,$$

$$P_{i+1} = a_i Q_i - P_i,$$

and

$$Q_{i+1} = (D - P_{i+1}^2)/Q_i.$$

Also compute G_i and B_i as follows. Begin with $G_{-2} = -P_0$, $G_{-1} = Q_0$, $B_{-2} = 1$, and $B_{-1} = 0$. Then for $i \geq 0$, set $G_i = a_i G_{i-1} + G_{i-2}$, and set $B_i = a_i B_{i-1} + B_{i-2}$. Sometimes one also computes A_i as $A_{-2} = 0$, $A_{-1} = 1$, and $A_i = a_i A_{i-1} + A_{i-2}$ for $i \geq 0$. Then $G_i = Q_0 A_i - P_0 B_i$.

Note that $G_i^2 - DB_i^2 = (-1)^{i+1} Q_{i+1} Q_0$. This relation will be important to us because all of the methods of solution we discuss will involve setting $Q_0 = |N|$, and finding those i so that $(-1)^{i+1} Q_{i+1} = N/|N|$. Then (G_i, B_i) will be a solution to the equation being considered. From a computational viewpoint, also note that, in some sense, G_i and B_i will typically be large, while Q_0 and Q_{i+1} will be small. So this equation sometimes allows accurate computation of the left-hand side when numbers on the left-hand side exceed the machine accuracy available. Exactly how far to carry these computations is discussed with each use below.

The sequence a_i is the simple continued fraction expansion of $(P_0 + \sqrt{D})/Q_0$, and the A_i/B_i are the convergents to this continued fraction. Each of the sequences P_i, Q_i, and a_i is periodic from some point, although not necessarily the same point for all three. Starting from the right point, the periodic part of the sequence P_i is palindromic. For each of the sequences Q_i and a_i, the periodic part, less the last term, is palindromic.

To solve the equation $x^2 - Dy^2 = \pm 1$, apply the PQa algorithm with $P_0 = 0$ and $Q_0 = 1$. There will be a smallest i with $a_i = 2a_0$, which will also be the smallest $i > 0$ so that $Q_i = 1$. There are two cases to consider: this i is odd, or this i is even.

If this i is odd, then the equation $x^2 - Dy^2 = -1$ has solutions. The minimal positive solution is given by $x = G_{i-1}, y = B_{i-1}$. For any positive integer k, if k is odd then $x = G_{ki-1}, y = B_{ki-1}$ is a solution to the equation $x^2 - Dy^2 = -1$, and all solutions to this equation with x and y positive are generated this way. If k is an even positive integer, then $x = G_{ki-1}, y = B_{ki-1}$ is a solution to the equation

$x^2 - Dy^2 = 1$, and all solutions to this equation with x and y positive are generated this way. The minimal positive solution to $x^2 - Dy^2 = 1$ is $x = G_{2i-1}, y = B_{2i-1}$.

If the smallest i so that $a_i = 2a_0$ is even, then the equation $x^2 - Dy^2 = -1$ does not have any solutions. For any positive integer k, $x = G_{ki-1}, y = B_{ki-1}$ is a solution to the equation $x^2 - Dy^2 = 1$, and all solutions to this equation with x and y positive are generated this way. In particular, the minimal positive solution to $x^2 - Dy^2 = 1$ is $x = G_{i-1}, y = B_{i-1}$.

The sequences P_j and a_j are periodic with period i after the zero-th term, i.e., the first period is P_1 to P_i for the sequences P_j, and a_1 to a_i for the sequence a_j. The sequence Q_j is periodic starting at the zero-th term, i.e., the first period is Q_0 to Q_{i-1}.

In Sections 3.2–3.5 and 3.6, respectively, we give several methods to generate all solutions to either Pell's and negative Pell's equations once the minimal positive solution is found.

4.4.2 Solving the Special Equations $x^2 - Dy^2 = \pm 4$

In some ways, solutions to the equation $x^2 - Dy^2 = \pm 4$ are more fundamental than solutions to the equation $x^2 - Dy^2 = \pm 1$. The most interesting case is when $D \equiv 1 \pmod 4$, so we cover that first.

When $D \equiv 1 \pmod 4$, apply the PQa algorithm with $P_0 = 1$ and $Q_0 = 2$. There will be a smallest $i > 0$ so that $a_i = 2a_0 - 1$. This will also be the smallest $i > 0$ so that $Q_i = 2$. The minimal positive solution to $x^2 - Dy^2 = \pm 4$ is then $x = G_{i-1}$, $y = B_{i-1}$. If i is odd, it will be a solution to the -4 equation, while if i is even it will be a solution to the $+4$ equation and the -4 equation will not have solutions. Periodicity of the sequences P_i, Q_i, and a_i is similar to that for the ± 1 equation.

If $D \equiv 0 \pmod 4$, then for any solution to $x^2 - Dy^2 = \pm 4$, x must be even. Set $X = x/2$, set $Y = y$, and solve $X^2 - (D/4)Y^2 = \pm 1$. If (X, Y) is the minimal positive solution to this equation, then $x = 2X, y = Y$ is the minimal positive solution to $x^2 - Dy^2 = \pm 4$. Alternatively, one can apply the PQa algorithm with $P_0 = 0$ and $Q_0 = 2$. If i is the smallest index so that $a_i = 2a_0$, then the minimal positive solution is (G_{i-1}, B_{i-1}).

If $D \equiv 2$ or $3 \pmod 4$, then by considerations modulo 4 one can that both x and y must be even. Set $X = x/2$, set $Y = y/2$, and solve $X^2 - DY^2 = \pm 1$. If (X, Y) is the minimal positive solution to this equation, then $x = 2X, y = 2Y$ is the minimal positive solution to $x^2 - Dy^2 = \pm 4$. Alternatively, use the PQa algorithm with $P_0 = 0$ and $Q_0 = 1$, but set $G_{-2} = 0$, $G_{-1} = 2$, $B_{-2} = 2$, and $B_{-1} = 0$. If i is the smallest index so that $a_i = 2a_0$, then the minimal positive solutions is (G_{i-1}, B_{i-1}).

As with the ± 1 equation, all solutions can be generated from the minimal positive solution. Consider first the equation $x^2 - Dy^2 = 4$. If (x_1, y_1) is the minimal positive solution to this equation, then for the n-th solution we have

$$x_n + y_n\sqrt{D} = \frac{1}{2^{n-1}}(x_1 + y_1\sqrt{D})^n$$

$$x_n - y_n\sqrt{D} = \frac{1}{2^{n-1}}(x_1 - y_1\sqrt{D})^n.$$

(4.4.1)

Therefore

$$x_n = \left(\frac{x_1 + y_1\sqrt{D}}{2}\right)^n + \left(\frac{x_1 - y_1\sqrt{D}}{2}\right)^n$$

$$y_n = \frac{1}{\sqrt{D}}\left[\left(\frac{x_1 + y_1\sqrt{D}}{2}\right)^n - \left(\frac{x_1 - y_1\sqrt{D}}{2}\right)^n\right].$$

(4.4.2)

We also have the recursion

$$x_{n+1} = \frac{1}{2}(x_1 x_n + D y_1 y_n)$$

$$y_{n+1} = \frac{1}{2}(y_1 x_n + x_1 y_n)$$

(4.4.3)

The relations (4.4.3) could be written in the following useful matrix form

$$\begin{pmatrix} x_{n+1} \\ y_{n+1} \end{pmatrix} = \frac{1}{2}\begin{pmatrix} x_1 & D y_1 \\ y_1 & x_1 \end{pmatrix}\begin{pmatrix} x_n \\ y_n \end{pmatrix}$$

(4.4.4)

from where

$$\begin{pmatrix} x_n \\ y_n \end{pmatrix} = \frac{1}{2^n}\begin{pmatrix} x_1 & D y_1 \\ y_1 & x_1 \end{pmatrix}^n\begin{pmatrix} x_0 \\ y_0 \end{pmatrix}$$

(4.4.5)

where $(x_0, y_0) = (2, 0)$ is the trivial solution.

We can express all integer solutions to the positive equation by the following formula

$$\frac{1}{2}(u_n + v_n\sqrt{D}) = \varepsilon_n\left(\frac{u_1 + v_1\sqrt{D}}{2}\right)^n, \quad n \in \mathbb{Z},$$

(4.4.6)

where ε_n is 1 or -1. Indeed, for $n > 0$ and $\varepsilon_n = 1$ we get all negative solutions. For $n > 0$ and $\varepsilon_n = -1$ we obtain all solutions (u_n, v_n) with u_n and v_n negative. For $n < 0$ and $\varepsilon_n = 1$ we have (u_n, v_n) with $u_n > 0$ and $v_n < 0$, while $n < 0$ and $\varepsilon_n = -1$ gives $u_n < 0$ and $v_n > 0$. The trivial solutions $(2, 0)$ and $(-2, 0)$ are obtained for $n = 0$.

Formula (4.4.6) captures all symmetries of equation $(u, v) \rightarrow (-u, -v)$, $(u, v) \rightarrow (u, -v)$, $(u, v) \rightarrow (-u, v)$. Therefore, in 2D the points (u_n, v_n) represents the orbits of the action of the Klein four-group $\mathbb{Z}_2 \times \mathbb{Z}_2$, i.e., points obtained by the 180 degree rotation, the vertical reflection, and by the horizontal reflection.

Now suppose the equation $x^2 - Dy^2 = -4$ has solutions, let (x_1, y_1) be the minimal positive solution, and define x_n, y_n by the equation $x_n + y_n\sqrt{D} = [(x_1 + y_1\sqrt{D})^n]/(2^{n-1})$. Then if n is odd, (x_n, y_n) is a solution to the equation $x^2 - Dy^2 = -4$, and if n is even then (x_n, y_n) is a solution to the equation $x^2 - Dy^2 = 4$. All positive solutions to these two equations are so generated. The pair (x_n, y_n) in (4.4.2) also alternately generates solutions to the $+4$ and -4 equation.

The set of solutions can be summarized as follows.

Theorem 4.4.1. *Let (x_1, y_1) be the minimal positive solution to $x^2 - Dy^2 = \pm 4$. Then for any solution to $x^2 - Dy^2 = \pm 4$, there is a choice of signs $+$ and $-$, and an integer n such that*

$$\frac{1}{2}(x + y\sqrt{D}) = \pm \left(\frac{x_1 + y_1\sqrt{D}}{2}\right)^n. \tag{4.4.7}$$

In some ways, the equation $x^2 - Dy^2 = \pm 4$ is more fundamental than the equation $x^2 - Dy^2 = \pm 1$. The numbers 1 and 4 are the only N's so that, for any D, if you know the minimal positive solution to the equation $x^2 - Dy^2 = \pm N$, you can generate all solutions, and you can do this without solving any other Pell's equation. Also, if you know the minimal positive solution to $x^2 - Dy^2 = \pm 4$, you can generate all the solutions to $x^2 - Dy^2 = \pm 1$. But the converse does not hold. The best that can be said as a converse is that for D not 5 or 12, the solutions to the equation $x^2 - Dy^2 = \pm 4$ can be derived from the intermediate steps when the PQa algorithm is used to solve the equation $x^2 - Dy^2 = \pm 1$.

When $D \equiv 1 \pmod 4$, considerations modulo 4 show that for any solution to $x^2 - Dy^2 = \pm 4$, x and y are both odd or both even. If the minimal positive solution has both x and y even, then all solutions have both x and y even. In this case, every solution to $x^2 - Dy^2 = \pm 1$ is just one-half of a solution to $x^2 - Dy^2 = \pm 4$. If the minimal positive solution to $x^2 - Dy^2 = \pm 4$ has both x and y odd, then $D \equiv 5 \pmod 8$, every third solution has x and y even, and all other solutions have x and y odd. In this case, every solution to $x^2 - Dy^2 = \pm 1$ is just one-half of one of the solutions to $x^2 - Dy^2 = \pm 4$ that has both x and y even. When $D \equiv 1 \pmod 4$, the equation $x^2 - Dy^2 = -4$ has solutions if and only if the equation $x^2 - Dy^2 = -1$ has solutions.

When $D \equiv 0 \pmod 4$, considerations modulo 4 show that for any solution to $x^2 - Dy^2 = \pm 4$, x is even. If the minimal positive solution has y even, then all solutions have y even (and x is always even). In this case, every solution to $x^2 - Dy^2 = \pm 1$ is just one-half of a solution to $x^2 - Dy^2 = \pm 4$. If the minimal positive solution to $x^2 - Dy^2 = \pm 4$ has y odd, then every other solution has y even, and every other solution has y odd. In this case, every solution to $x^2 - Dy^2 = \pm 1$ is just one-half of one of the solutions to $x^2 - Dy^2 = \pm 4$ that has x and y both even. When $D \equiv 0 \pmod 4$, it is possible for these to be solutions to $x^2 - Dy^2 = -4$, but not solutions to $x^2 - Dy^2 = 1$. This happens for $D = 8, 20, 40, 52$ and many more values. Of course, $x^2 - Dy^2 = -1$ never has solutions when $D \equiv 0 \pmod 4$.

When $D \equiv 2 \pmod 4$ or $D \equiv 3 \pmod 4$, all solutions to $x^2 - Dy^2 = \pm 4$ have both x and y even. Every solution to $x^2 - Dy^2 = \pm 1$ is just one-half of a solution to $x^2 - Dy^2 = \pm 4$. The equation $x^2 - Dy^2 = -4$ has solutions if and only if the equation $x^2 - Dy^2 = -1$ has solutions.

The cases $D \equiv 1 \pmod 4$ for D squarefree, and $D = 4r$ for $r \equiv 2$ or $3 \pmod 4$, r squarefree, are treated in [53]. The material we have presented above is not addressed directly in either [159] or [142]. For example, the proof that the method for solving the equation works in the case $d \equiv 1 \pmod 4$ is not trivially derived from the material in one or both of these sources.

Remark. Concerning the equation $x^2 - Dy^2 = -4$ the following conjecture is still open: *Let p be a prime $\equiv 1 \pmod 4$ and let (x_1, y_1) be the fundamental solution to the equation $x^2 - py^2 = -4$. Then $y_1 \not\equiv 0 \pmod p$.*

This has been verified for all primes $p < 2000$ with $p \equiv 5 \pmod 8$ and for all primes $p < 100000$ with $p \equiv 1 \pmod 8$. Also, it has been shown that $y_1 \not\equiv 0 \pmod p$ if and only if $B_{\frac{p-1}{4}} \not\equiv 0 \pmod p$, where the Bernoulli numbers B_n are defined by the series

$$\frac{t}{e^t - 1} = 1 - \frac{t}{2} + \sum_{n=1}^{\infty} \frac{(-1)^{n-1} B_n}{(2n)!} t^{2n}.$$

4.4.3   Structure of Solutions to the General Pell's Equation

As we have seen in Section 4.1, if (r, s) is a solution to $x^2 - Dy^2 = N$, and (t, u) is any solution to its Pell's resolvent, then for $x = rt + Dsu$, $y = ru + st$, (x, y) is a solution to the equation $x^2 - Dy^2 = N$. This follows from the multiplication principle:

$$(r^2 - Ds^2)(t^2 - Du^2) = (rt + Dsu)^2 - D(ru + st)^2.$$

This fact can be used to separate solutions to $x^2 - Dy^2 = N$ into equivalence classes. Two solutions (x, y) and (r, s) are equivalent if there is a solution (t, u) to $t^2 - Du^2 = 1$ so that $x = rt + Dsu$ and $y = ru + st$.

It may help to view the set of solutions geometrically. If $N > 0$, then, as an equation in real numbers, $x^2 - Dy^2 = N$ is a hyperbola with the x-axis as its axis, and the y-axis as an axis of symmetry. The asymptotes are the lines $x \pm y\sqrt{D} = 0$. Let (t, u) be the minimal positive solution to $x^2 - Dy^2 = 1$. Draw the graph of $x^2 - Dy^2 = N$ over the reals. Mark the point $(\sqrt{N}, 0)$, which is on this graph.

Now mark the point $(t\sqrt{N}, u\sqrt{N})$, which is also on the graph. Continue marking points so that if (x, y) is the most recent point marked, then the next point marked is $(xt + Dyu, xu + yt)$. All of the points marked so far, apart from the first, have $x > 0$ and $y > 0$. Now, for each point (x, y) that has been marked, mark all of the points $(\pm x, \pm y)$ not yet marked.

The marked points divide the graph into intervals. Make the interval $((\sqrt{N}, 0), (t\sqrt{N}, u\sqrt{N})]$ a half-open interval, and then make the other intervals on this branch half-open by assigning endpoints to one interval. Make the intervals on the other half-open by mapping (x, y) in the right branch to $(-x, -y)$ on the left branch. If there are integer solutions to $x^2 - Dy^2 = N$, then

1) No two solutions within the same (half-open) interval are equivalent,
2) Every interval has exactly one solution in each class, and
3) The order of solutions by class is the same in every interval.

Instead of starting with the point $(\sqrt{N}, 0)$, we could have started with any point (r, s) on the graph, and marked off the points corresponding to $\pm(r + s\sqrt{D})(t + u\sqrt{D})^n$. The above three comments still apply.

The situation is similar in the case $N < 0$, except that the graph has the y-axis as its axis, and the x-axis is an axis of symmetry. If the negative Pell's equation $x^2 - Dy^2 = -1$ is solvable, then any of its solutions can be used to generate a correspondence between solutions to $x^2 - Dy^2 = N$ and $x^2 - Dy^2 = -N$.

Within a class there is a unique solution with x and y nonnegative, but smaller than any nonnegative solution. This is the minimal nonnegative solution for the class. There is also either one or two solutions so that y is nonnegative, and is less than or equal to any other nonnegative y in any solution (x, y) within the class. If there is one such solution, it is called the fundamental solution. If there are two such solutions, then they will be equivalent and their x-values will be negatives of each other. In this case, the solution with the positive x-value is called the fundamental solution for the class.

When tabulating solutions, it is usually convenient to make a list consisting of one solution from each class. Often, this list will consist of the minimal nonnegative solutions, or the fundamental solutions. Given any solution in a class, it is easy to find the fundamental solution or the minimal nonnegative solution for that class.

The results are summarized in the following

Theorem 4.4.2. *Given any solution in a class, all solutions in that class are derived from solutions to the equation $x^2 - Dy^2 = 1$. If (r, s) is any particular solution to $x^2 - Dy^2 = N$, (x, y) is any other solution to the same equation in the same class as (r, s) and if (t_1, u_1) is the fundamental solution to the Pell's resolvent, then for some choice of signs $+$ and $-$, and for some integer n*

$$x + y\sqrt{D} = \pm(r + s\sqrt{D})(t_1 + u_1\sqrt{D})^n. \tag{4.4.8}$$

We can write formulas similar to those presented for cases $N = \pm 1$ and $N = \pm 4$ in Section 4.4.2.

4.4.4   Solving the Equation $x^2 - Dy^2 = N$ for $N < \sqrt{D}$

When $1 < N^2 < D$, apply the PQa algorithm with $P_0 = 0$, $Q_0 = 1$. Continue the computations until you reach the first $i > 0$ with $G_i^2 - DB_i^2 = 1$ (i.e., $Q_{i+1} = 1$ and $i + 1$ is even). For $1 \le j \le i$, if $G_j^2 - DB_j^2 = N/f^2$ for some $f > 0$, add fG_j, fB_j to the list of solutions. When done, the list of solutions will have the minimal positive member of each class.

The list of all solutions can be generated using the methods of the previous section. Alternatively, all positive solutions can be generated by extending the PQa algorithm indefinitely.

4.4.5   Solving the Equation $x^2 - Dy^2 = N$ by Brute-Force Search

Let (t, u) be the minimal positive solution to $x^2 - Dy^2 = N$. If $N > 0$, set $y_1 = 0$, and $y_2 = \sqrt{\dfrac{(t-1)N}{2D}}$. If $N < 0$, set $y_1 = \sqrt{\dfrac{|N|}{2}}$, and $y_2 = \sqrt{\dfrac{(t+1)|N|}{2D}}$. For $y_1 \le y \le y_2$, if $N + Dy^2$ is a square, set $x = \sqrt{N + Dy^2}$. If (x, y) is not equivalent to $(-x, y)$, add both to the list of solutions, otherwise just add (x, y) to the list. When finished, this list gives the fundamental solutions.

This method works well if y_2 is not too large, which means that $\sqrt{\dfrac{(t\pm1)|N|}{2D}}$ is not too large. Hence it suffices to search between the bounds y_1 and y_2.

To generate all solutions be performing this algorithm, refer to the structure of solutions of general Pell's equation given in Section 4.4.3.

4.4.6   Numerical Examples

In order to see how the algorithms that we have presented work, we will examine a few numerical examples. Computations were done by using MATHEMATICA.

Example 1. Consider the equations

$$x^2 - 109y^2 = \pm 1.$$

Apply the PQa algorithm with $P_0 = 0$ and $Q_0 = 1$. The following table gives the index, i, and then the several calculated quantities for $i = -2$ to 30.

i	P_i	Q_i	a_i	G_i	B_i	$G^2 - 109B^2$
-2				0	1	-109
-1				1	0	1
0	0	1	10	10	1	-9
1	10	9	2	21	2	5
2	8	5	3	73	7	-12
3	7	12	1	94	9	7
4	5	7	2	261	25	-4
5	9	4	4	1138	109	15
6	7	15	1	1399	134	-3
7	8	3	6	9532	913	3
8	10	3	6	58591	5612	-15
9	8	15	1	68123	6525	4
10	7	4	4	331083	31712	-7
11	9	7	2	730289	69949	12
12	5	12	1	1061372	101661	-5
13	7	5	3	3914405	374932	9
14	8	9	2	8890182	851525	-1
15	10	1	20	181718045	17405432	9
16	10	9	2	372326272	35662389	-5
17	8	5	3	1298696861	124392599	12
18	7	12	1	1671023133	160054988	-7
19	5	7	2	4640743127	444502575	4
20	9	4	4	20233995641	1938065288	-15
21	7	15	1	24874738768	2382567863	3
22	8	3	6	169482428249	16233472466	-3
23	10	3	6	1041769308262	99783402659	15
24	8	15	1	1211251736511	116016875125	-4
25	7	4	4	5886776254306	563850903159	7
26	9	7	2	12984804245123	1243718681443	-12
27	5	12	1	18871580499429	18075659584602	5
28	7	5	3	69599545743410	6666427435249	-9
29	8	9	2	158070671986249	15140424455100	1
30	10	1	20	3231012985468390	309474916537249	-9

We have $a_0 = 10$, and the first i so that $a_i = 2a_0$ is $i = 15$, at which point $a_{15} = 20$. Hence the period of a_i is 15, which is odd, and so the equation $x^2 - 109y^2 = -1$ has solutions. The minimal positive solution to $x^2 - 109y^2 = -1$ is $x = 8890182$, $y = 851525$. The minimal positive solution to $x^2 - 109y^2 = 1$ is $x = 158070671986249$, $y = 15140424455100$.

Example 2. Let us examine now the equations

$$x^2 - 109y^2 = \pm 4.$$

As $D \equiv 1 \pmod 4$, apply the PQa algorithm with $P_0 = 1$ and $Q_0 = 2$. The following table gives the index, i, and the standard quantities for $i = -2$ to 14.

i	P_i	Q_i	a_i	G_i	B_i	$G^2 - 109B^2$
-2				-1	1	-108
-1				2	0	4
0	1	2	5	9	1	-28
1	9	14	1	11	1	12
2	5	6	2	31	3	-20
3	7	10	1	42	4	20
4	3	10	1	73	7	-12
5	7	6	2	188	18	28
6	5	14	1	261	25	-4
7	9	2	9	2537	243	28
8	9	14	1	2798	268	-12
9	5	6	2	8133	779	20
10	7	10	1	10931	1047	-20
11	3	10	1	19064	1826	12
12	7	6	2	49059	4699	-28
13	5	14	1	68123	6535	4
14	9	2	9	662166	63424	-28

We have $a_0 = 5$, and the first i so that $a_i = 2a_0 - 1$ is $i = 7$, at which point $a_7 = 9$. Hence the period of a_i is 7, which is odd, and so the equation $x^2 - 109y^2 = -4$ has solutions. The minimal positive solution to $x^2 - 109y^2 = -4$ is $x = 261$, $y = 25$. The minimal positive solution to $x^2 - 109y^2 = 4$ is $x = 68123$, $y = 6525$.

Note that the third solution to $x^2 - 109y^2 = \pm 4$ can be computed from

$$\frac{1}{4}(261 + 25\sqrt{109})^3 = 17780364 + 1703050\sqrt{109}.$$

Upon dividing by 2, we get the minimal positive solution $x = 8890182$, $y = 851525$ to negative Pell's equation $x^2 - 109y^2 = -1$.

Example 3. Consider the equation

$$x^2 - 129y^2 = -5.$$

From Theorem 4.1.4 it follows that if this equation is solvable, then it has exactly two classes of solutions.
Here $N < \sqrt{D}$. Apply the PQa algorithm with $P_0 = 0$, $Q_0 = 1$.

i	P_i	Q_i	a_i	G_i	B_i	$G^2 - 129B^2$
-2				0	1	-129
-1				1	0	1
0	0	1	11	11	1	-8
1	11	8	2	23	2	13
2	5	13	1	34	3	-5
3	8	5	3	125	11	16
4	7	16	1	159	14	-3
5	9	3	6	1079	95	16
6	9	16	1	1238	109	-5
7	7	5	3	4793	422	13
8	8	13	1	6031	531	-8
9	5	8	2	16855	1484	1
10	11	1	22	376841	33179	-8

The only $f > 0$ so that f^2 divides -5 is $f = 1$. Reviewing the above for $G_j^2 - 129B_j^2 = -5$, we find solutions (x, y) equal to (34,3) and (1238,109). Thus, there are two classes of solutions, and these are the minimal positive solutions for these classes.

Example 4. Let us use the brute-force search method to find the fundamental solutions of

$$x^2 - 61y^2 = 15.$$

The minimal positive solution to Pell's resolvent $x^2 - 61y^2 = 1$ is $x = 1766319049$, $y = 226153980$. As $N = 15$ is positive, the lower search limit for y is 0, and the upper limit is

$$\sqrt{\frac{15(1766319049 - 1)}{2 \cdot 61}} \approx 14736, 702.$$

So we search on y from 0 to 14736. Only $y = 11$ and $y = 917$ yield solutions, so the four fundamental solutions are $x = \pm 86$, $y = 11$, and $x = \pm 7162$, $y = 917$.

Example 5. For the same equation above, we will apply now the LMM algorithm given in Section 4.3.

The only $f > 0$ so that f^2 divides 15 is $f = 1$. Set $m = 15$. The z's with $-15/2 < z \le 15/2$ and $z^2 \equiv 61 \pmod{15}$ are $z = \pm 1$, $z = \pm 4$.

Upon performing the PQa algorithm with $P_0 = 1$, $Q_0 = 15$, and $d = 61$, the first $Q_i = \pm 1$ occurs at $Q_9 = 1$. The corresponding solution has $x = G_8 = 2593$ and $y = B_8 = 332$. For this (x, y), $x^2 - 61y^2 = -15$. The equation $x^2 - 61y^2 = -1$ is solvable and the minimal positive solution is $x = 29718$, $y = 3805$. Applying this to the solution (2593, 332) gives the solution $x = 154117634$, $y = 19732741$ to the equation $x^2 - 61y^2 = 15$. This is equivalent to the fundamental solution $(-86, 11)$.

Performing the PQa algorithm with $P_0 = -1$, $Q_0 = 15$, and $d = 61$, gives the first $Q_i = \pm 1$ at $Q_4 = 1$, yielding the fundamental solution $(86, 11)$ to the equation $x^2 - 61y^2 = 15$.

Performing the PQa algorithm with $P_0 = 4$, $Q_0 = 15$, and $d = 61$, gives the first $Q_i = \pm 1$ ate $Q_{10} = 1$, yielding the fundamental solution $(7162, 917)$ to the equation $x^2 - 61y^2 = 15$.

Performing the PQa algorithm with $P_0 = -4$, $Q_0 = 15$, and $d = 61$, gives the first $Q_i = \pm 1$ at $Q_3 = 1$, yielding the solution $(31, 4)$ to the equation $x^2 - 61y^2 = -15$. Applying the minimal positive solution to $x^2 - 61y^2 = -1$ gives the solution $x = 1849678$, $y = 236827$ to the equation $x^2 - 61y^2 = 15$. This is equivalent to the fundamental solution $(-7162, 917)$.

4.5 Solvability and Unsolvability of the Equation $ax^2 - by^2 = c$

Using the results in our papers [13, 14] and [17] we will present two general methods for solving the equation

$$ax^2 - by^2 = c. \tag{4.5.1}$$

We will also give sufficient conditions such that equation (4.5.1) is unsolvable in positive integers. In the special case $c = 1$, the equation (4.5.1) was studied in Section 3.5.

The equation (4.5.1) is also considered in the recent paper [114], where a, b, c are positive integers with $\gcd(a, b) = 1$. The author showed that if (4.5.1) is solvable, then it has infinitely many positive integer solutions. But his result is in fact a variant of the multiplication principle.

The following result given in [172] completely solves the problem of determining all solutions to equation (4.5.1).

Theorem 4.5.1. *Let a, b be positive integers such that $\gcd(a, b) = 1$ and a is squarefree, and let c be a nonzero integer. Denote $D = ab$, $N = ac$. Then (u, v) is a solution to the general Pell's equation*

$$u^2 - Dv^2 = N \tag{4.5.2}$$

if and only if $\left(\dfrac{u}{a}, v\right)$ is solution to (4.5.1).

Proof. Let (x, y) be a solution to (4.5.1). It follows that $(ax)^2 - aby^2 = ac$, so (ax, y) is a solution to the associated general Pell's equation (4.5.2).

Conversely, if (u, v) is a solution to (4.5.2), from the relation $u^2 - abv^2 = ac$ we obtain $a \mid u^2$. Taking into account that a is squarefree it follows that $a \mid u$. Therefore $u = a_1 a$ and $(a_1 a)^2 - abv^2 = ac$ yield $aa_1^2 - bv^2 = c$, i.e., $\left(\dfrac{u}{a}, v\right)$ is a solution to (4.5.1).   □

Remarks. 1) From the above result it is clear that (4.5.1) is solvable if and only if the associated general Pell's equation (4.5.2) is solvable.

2) The assumption that a is squarefree is not a restriction. Indeed, if $a = a_1 m^2$ and a_1 is squarefree, then the equation (4.5.1) becomes $a_1 X^2 - by^2 = c$, where $X = mx$, i.e., an equation of the same type.

3) In order to solve (4.5.1) we determine all solutions (u, v) to the general Pell's equation (4.5.2). The desired solutions are given by $\left(\dfrac{u}{a}, v\right)$.

The equation (4.5.1) is strongly connected to the general Pell's equation (4.5.2) and to the Diophantine equation

$$as^2 - bt^2 = 1. \tag{4.5.3}$$

The solvability of these three equations is studied in the following theorem [17]:

Theorem 4.5.2. *Suppose that* $\gcd(a, b) = 1$ *and ab is not a perfect square. Then:*

1) If the equations (4.5.2) and (4.5.3) are solvable, then (4.5.1) is also solvable and all of its solutions (x, y) *are given by*

$$x = s_0 u + b t_0 v, \quad y = t_0 u + a s_0 v \qquad (4.5.4)$$

where (u, v) *is any solution to (4.5.2) and* (s_0, t_0) *is the minimal solution to (4.5.3).*

2) If the equations (4.5.1) and (4.5.3) are solvable, then (4.5.2) is also solvable.

3) If the equations (4.5.1) and (4.5.2) are solvable and there exist solutions (x, y), (u, v) *such that*

$$\frac{ux - bvy}{c} \quad and \quad \frac{-avx + uy}{c}$$

are both integers, then (4.5.3) is solvable.

Proof. 1) We have

$$ax^2 - by^2 = a(s_0 u + b t_0 v)^2 - b(t_0 u + a s_0 v)^2 =$$
$$= (a s_0^2 - b t_0^2)(u^2 - abv^2) = 1 \cdot c = c,$$

and it follows that (x, y), given in (4.5.4), is a solution to the equation (4.5.1).

Conversely, let (x, y) be a solution to (4.5.1), and let (s_0, t_0) be the minimal solution to the equation (4.5.3). Then (u, v), where $u = a s_0 x - b t_0 y$ and $v = -t_0 x + s_0 y$ is a solution to the general Pell's equation (4.5.2). Solving the above system of linear equations with unknowns x and y yields $x = s_0 u + b t_0 v$ and $y = t_0 u + a s_0 v$, i.e., (x, y) has the form (4.5.4).

2) If (x, y) and (s, t) are solutions to (4.5.1) and (4.5.3), respectively, then (u, v), with $u = asx - bty$ and $v = -tx + sy$ is a solution to (4.5.2). Moreover, each solution to (4.5.2) is of the above form. Indeed, if (u, v) is an arbitrary solution to (4.5.2), then (x, y), where $x = su + btv$ and $y = tu + asv$, is a solution to (4.5.1). Thus, solving the above system of linear equations in u, v, it follows that $u = asx - bty$ and $v = -tx + sy$.

3) Let (x, y) and (u, v) be solutions to (4.5.1) and (4.5.2), respectively, for which

$$s = \frac{ux - bvy}{c} \in \mathbb{Z} \quad and \quad t = \frac{-avx + uy}{c} \in \mathbb{Z}.$$

Then (s, t) is a solution to (4.5.3). □

Remarks. 1) The equation $8x^2 - y^2 = 7$ is solvable and all of its solutions were determined in Section 4.1. For this equation, the associated equations (4.5.2) and (4.5.3) are $u^2 - 8v^2 = 7$ and $8s^2 - t^2 = 1$, respectively. It is interesting to see that both these equations are unsolvable.

2) In case of solvability of equations (4.5.2) and (4.5.3), the formulas (4.5.4) point
out an alternative way to express the solutions to equation (4.5.1).

Theorem 4.5.3. *Let a, c be relatively prime positive integers, not both perfect squares, and let b and d be integers. The equation*

$$ax^2 - cy^2 = ad - bc \tag{4.5.5}$$

is solvable if and only if the numbers $an + b$ and $cn + d$ are perfect squares for some positive integer n. In this case, the number of such n's is infinite.

Proof. If (x_0, y_0) is a solution to the equation (4.5.5), then by Theorem 4.5.2, $(x_m, y_m)_{m \geq 0}$, where

$$x_m = x_0 u_m + c y_0 v_m, \quad y_m = a x_0 v_m + y_0 u_m \tag{4.5.6}$$

are solutions to this equation. Here $(u_m, v_m)_{m \geq 0}$ denotes the general solution to Pell's equation $u^2 - acv^2 = 1$.

Then $ax_m^2 - cy_m^2 = ad - bc$, $m = 0, 1, 2, \ldots$, hence

$$a(x_m^2 - d) = c(y_m^2 - b), \quad m = 0, 1, 2, \ldots \tag{4.5.7}$$

Since a and c are relatively prime, from (4.5.7) it follows that $a \mid y_m^2 - b$ and $c \mid x_m^2 - d$, $m = 0, 1, 2, \ldots$. Let

$$n_m = \frac{y_m^2 - b}{a} = \frac{x_m^2 - d}{c}, \quad m = 0, 1, 2, \ldots \tag{4.5.8}$$

Clearly, n_m is a positive integer for each m and

$$a n_m + b = y_m^2, \quad c n_m + d = x_m^2, \quad m = 0, 1, 2, \ldots$$

i.e., the numbers $an + b$ and $cn + d$ are simultaneously perfect squares for infinitely many positive integers n.

If the equation (4.5.5) is not solvable in positive integers, then $an + b$ and $cn + d$ cannot be both perfect squares. Indeed, if we assume that there is a positive integer n_0 such that $an_0 + b = y_0^2$ and $cn_0 + d = x_0^2$ for some positive integers x_0, y_0, then by eliminating n_0 it follows that $ax_0^2 - by_0^2 = ad - bc$, in contradiction with the unsolvability of equation (4.5.5).   □

Theorem 4.5.4. *Let a and b be positive integers such that for all positive integers n, $an + b$ is not a perfect square. Then the equations*

$$ax^2 - (am + v_0)y^2 = c, \quad m = 0, 1, 2, \ldots \tag{4.5.9}$$

and

$$(am + w_0)x^2 - ay^2 = c, \quad m = 0, 1, 2, \ldots \tag{4.5.10}$$

are not solvable in positive integers. Here c is a nonzero integer and (u_0, v_0) and (w_0, s_0) are the minimal solutions to the equations $au - bv = c$ and $bw - as = c$, respectively.

Proof. The general solutions to the linear Diophantine equations $au - bv = c$ and $bw - as = c$ are $(u_m, v_m)_{m \geq 0}$ and $(w_m, s_m)_{m \geq 0}$, respectively, where

$$u_m = u_0 + bm, \quad v_m = v_0 + am \quad \text{and} \quad w_m = w_0 + am, \quad s_m = s_0 + bm$$

(see [198]). Assume now that equation (4.5.9) is solvable and let (x, y) be a solution. Then $ax^2 - (am + v_0)y^2 = c$. But by considerations above, $c = au_m - bv_m$. It follows that

$$ax^2 - (am + v_0)y^2 = au_m - bv_m,$$

hence the equation (4.5.3), where $d = u_m$ and $c = u_m$ is solvable. From Theorem 4.5.3 we obtain that $an + b$ is a perfect square for some n, in contradiction with the hypothesis.   □

Example 1. The numbers $10n + 3$ are not perfect squares, $n = 0, 1, 2, \ldots$. Choosing $c = 1$, we find the minimal solutions to the equations $10u - 3v = 1$ and $3w - 10s = 1$. They are $(1, 3)$ and $(7, 2)$, respectively. From Theorem 4.5.4 it follows that the equations

$$10x^2 - (10m + 3)y^2 = 1 \quad \text{and} \quad (10m + 7)x^2 - 10y^2 = 1, \quad m = 0, 1, 2, \ldots$$

are not solvable.

Remark. In many situations it is not easy to find the minimal solutions (u_0, v_0) and (w_0, s_0) to the equations $au - bv = c$ and $bw - as = c$, respectively. In this cases we may replace (u_0, v_0) and (w_0, s_0) by any solution to the above equations and the results in Theorem 4.5.4 remain true.

Example 2. The numbers $5n + 2$ are not perfect squares for any positive integer n. The equations $5u - 2v = c$ and $2w - 5s = c$ have $(c, 2c)$ and $(3c, c)$ among their solutions, respectively. It follows that $u_m = c + 2m$, $v_m = 2c + 5m$, and $w_m = 3c + 5m$, $s_m = c + 2m$, $m = 0, 1, 2, \ldots$.
   From Theorem 4.5.3 we obtain that the equations

$$5x^2 - (5m + 2c)y^2 = c \quad \text{and} \quad (5m + 3c)x^2 - 5y^2 = c, \quad m = 0, 1, 2, \ldots$$

are not solvable.

Example 3. In a similar manner, starting with the nonsquare numbers $3n + 2$, $n = 0, 1, 2, \ldots$, we deduce that equations

$$3x^2 - (3m + c)y^2 = c \quad \text{and} \quad (3m + 2c)x^2 - 3y^2 = c, \quad m = 0, 1, 2, \ldots$$

are not solvable in positive integers.

There are many situations in which the unsolvability of an equation of the type (4.5.1) can be proven by using modular arithmetics arguments.

Example 4 ([193]). The equation

$$(4m + 3)x^2 - (4n + 1)y^2 = 1$$

where m and n are positive integers, is not solvable.

Indeed, $x^2, y^2 \equiv 0$ or $1 \pmod 4$ and so $(4m + 3)x^2 \equiv 0$ or $3 \pmod 4$ and $(4n + 1)y^2 \equiv 0$ or $1 \pmod 4$. By combining the residues, we obtain

$$(4m + 3)x^2 - (4n + 1)y^2 \not\equiv 1 \pmod 4.$$

Example 5 ([192]). In a similar manner, we can prove that equations

$$(4k + 2)x^2 - (4l + 3)y^2 = 1 \quad \text{and} \quad 7mx^2 - (7n + 1)y^2 = 1,$$

where k, l and m, n are positive integers, are also not solvable.

A criterion for solvability (unsolvability) for a class of general Pell's equations is given in [100] (see also subsection 4.2.4).

Theorem 4.5.5. *For N a squarefree integer, the equation*

$$x^2 - 2y^2 = N \tag{4.5.11}$$

is solvable if and only if it is solvable modulo N.

Proof. By multiplicativity, it suffices to show that $x^2 - 2y^2 = N$ has a solution for $N = -1, N = 2$, and $N = p$ for p an odd prime such that 2 is congruent to a square modulo p. For $N = -1$, use $1^2 - 2 \cdot 1^2 = -1$; for $N = 2$, use $2^2 - 2 \cdot 1^2 = 2$.

Now suppose p is an odd prime such that 2 is congruent to a square modulo p. Find x, y such that $x^2 - 2y^2$ is divisible by p but not by p^2 (if it is divisible by p^2, fix that by replacing x with $x + p$). Now form the ideal $(x + y\sqrt{D}, p)$. Its norm divides p^2 and $x^2 - 2y^2$, so it must be p.   □

Incidentally, one can replace 2 by any integer D such that $\mathbb{Q}(\sqrt{D})$ has unique factorization, provided that $x^2 - Dy^2 = -1$ has a solution. It turns out (but is by no means obvious!) that unique factorization implies that D is prime, and it is believed (but not proved) that $\mathbb{Q}(\sqrt{D})$ has unique factorization for about 75 % of the primes D. Moreover, existence of a solution of $x^2 - Dy^2 = -1$ then implies $D \equiv 1 \pmod 4$, but nor every prime congruent to 1 modulo 4 will work (try $D = 5$).

4.6 Solving the General Pell Equation by Using Quadratic Rings

The main purpose of this section is to present an algorithm for finding all positive integer solutions to the general Pell's equation

$$x^2 - Dy^2 = k \qquad (4.6.1)$$

where d is a nonsquare positive integer and k is a nonzero integer. We will follow the method described in [76] (see also [171]).

Let (x, y) be an integral solution of (4.6.1), i.e., $x^2 - Dy^2 = k$. We are going to use the results in Section 2.2.2. We have $N(\mu) = k$, where $\mu = x + \sqrt{D}y \in R$. If ε_0 is the fundamental unit of the ring R found in Theorem 3.4.1, then we will denote

$$\varepsilon = \begin{cases} \varepsilon_0, & \text{if } N(\varepsilon_0) = 1 \\ \varepsilon_0^2, & \text{if } N(\varepsilon_0) = -1. \end{cases}$$

Then the vectors $(1, 1)$ and $l(\varepsilon)$ form a base in the linear space $\mathbb{R}^2$. Indeed, if $\alpha(1, 1) + \beta l(\varepsilon) = 0$, with $\alpha, \beta \in \mathbb{R}$, then $\alpha + \beta \ln |\varepsilon| = 0$ and $\alpha + \beta \ln |\bar{\varepsilon}| = 0$. Since $\ln |\bar{\varepsilon}| = -\ln |\varepsilon| \neq 0$, from the previous two relations it follows that $\alpha = \beta = 0$.

If $\mu = x + y\sqrt{D} \in R$ and $N(\mu) = k$, then $k \neq 0$ implies $\mu \neq 0$, i.e., the vector $l(\mu)$ is well defined in $\mathbb{R}^2$. By using the fact that $(1, 1)$ and $l(\varepsilon)$ form a base in $\mathbb{R}^2$, we deduce the existence of $\alpha, \gamma \in \mathbb{R}$ such that $l(\mu) = \alpha(1, 1) + \gamma l(\varepsilon)$. This means that

$$\ln \mu = \alpha + \gamma \ln |\varepsilon| \quad \text{and} \quad \ln \bar{\mu} = \alpha + \gamma \ln |\bar{\varepsilon}|.$$

In particular, it follows that

$$\ln |k| = \ln |N(\mu)| = \ln |\mu| + \ln |\bar{\mu}| = 2\alpha + \gamma \ln |N(\varepsilon)| = 2\alpha,$$

i.e.,

$$\alpha = \frac{\ln |k|}{2} \quad \text{and} \quad l(\mu) = \frac{\ln |k|}{2}(1, 1) + \gamma l(\varepsilon).$$

Let a be the closest integer to γ, and let $\mu_0 = \varepsilon^{-a}\mu$. Then $\mu \sim \mu_0$ and $N(\mu_0) = N(\mu) = k$. In addition,

$$l(\mu_0) = \frac{\ln |k|}{2}(1, 1) + \gamma_1 l(\varepsilon),$$

where $|\gamma_1| \leq \dfrac{1}{2}$ and $\gamma_1 = \gamma - a$. Therefore

$$\ln |\mu_0| = \frac{\ln |k|}{2} + \gamma_1 \ln \varepsilon \quad \text{and} \quad \ln |\overline{\mu}_0| = \frac{\ln |k|}{2} + \gamma_1 \ln |\overline{\varepsilon}| = \frac{\ln |k|}{2} - \gamma_1 \ln \varepsilon$$

(we have used here that $\varepsilon > 1$). It follows that

$$\left| \ln |\mu_0| - \frac{\ln |k|}{2} \right| \leq \frac{1}{2} \ln \varepsilon \quad \text{and} \quad \left| \ln |\overline{\mu}_0| - \frac{\ln |k|}{2} \right| \leq \frac{1}{2} \ln \varepsilon.$$

The above inequalities can be written as

$$\ln \sqrt{\frac{|k|}{\varepsilon}} \leq \ln |\mu_0| \leq \ln \sqrt{\varepsilon |k|} \quad \text{and} \quad \ln \sqrt{\frac{|k|}{\varepsilon}} \leq \ln |\overline{\mu}_0| \leq \ln \sqrt{\varepsilon |k|}.$$

We obtain

$$\sqrt{\frac{|k|}{\varepsilon}} \leq |\mu_0| \leq \sqrt{\varepsilon |k|} \quad \text{and} \quad \sqrt{\frac{|k|}{\varepsilon}} \leq |\overline{\mu}_0| \leq \sqrt{\varepsilon |k|}. \tag{4.6.2}$$

The numbers $|\mu_0|$ and $|\overline{\mu}_0|$ can be written as $s + t\sqrt{D}$, where s and t are positive integers. Since $t\sqrt{D} \leq \max\{|\mu_0|, |\overline{\mu}_0|\} \leq \sqrt{\varepsilon |k|}$, we have

$$t \leq \sqrt{\frac{\varepsilon |k|}{D}} \quad \text{and} \quad s \leq \sqrt{\varepsilon |k|}. \tag{4.6.3}$$

We will now describe the actual algorithm.

Step 1. Search for elements $\mu_1, \mu_2, \ldots, \mu_r$ in R of the form $s + t\sqrt{D}$ such that s, t are positive integers satisfying inequalities (4.6.3) and $N(\mu_i) = k$, $i = 1, 2, \ldots, r$.

From the inequalities (4.6.3) it follows that there are finitely many such μ's in R. This fact also follows from Theorem 2.2.3.

Step 2. From Theorem 3.4.1 it follows that all elements $\mu \in R$ with $N(\mu) = k$ are of the form $\mu = \pm \mu_i \varepsilon^l$ or $\mu = \pm \overline{\mu}_i \varepsilon^l$, for some $i \in \{1, 2, \ldots, r\}$ and some integer l.

Finally, let us mention that we can determine the fundamental unit $\varepsilon_0 \in R$ in a finite number of steps. For this part we refer to Section 3.3, where we employed continued fractions.

4.7 Another Algorithm for Solving General Pell's Equation

In what follows we will present a different algorithm for solving the general Pell's equation (4.6.1). Our approach is based on the one given in [171] and [95].

It suffices to consider solutions (x, y) to (4.6.1) such that the positive integers x and y are relatively prime.

If $|k| < \sqrt{D}$, then we apply Theorem 3.3.1. When $k \neq (-1)^{n-1}q_{n+1}$ for all n, the equation (4.6.1) is not solvable. When $k = (-1)^{n-1}q_{n+1}$ for some n, the pair (h_n, k_n) is a solution to the general Pell's equation (4.6.1) and all other of its integral solutions are given by

$$x + y\sqrt{D} = (\pm h_n \pm k_n\sqrt{D})\varepsilon^l, \quad l \in \mathbb{Z},$$

where is the fundamental solution of the Pell's resolvent.

If $|k| > \sqrt{D}$, then we write $k = \delta k_0$, where $\delta = \pm 1$ and k_0 is a positive integer. Since x and y are relatively prime, there exist integers x_1 and y_1 such that $xy_1 - yx_1 = \delta$.

It follows that

$$(xx_1 - Dyy_1)^2 - D = (xx_1 - Dyy_1)^2 - D(xy_1 - yx_1)^2 =$$
$$= (x^2 - Dy^2)(x_1^2 - Dy_1^2) = k(x_1^2 - Dy_1^2) = \delta k_0(x_1^2 - Dy_1^2).$$

Hence

$$(xx_1 - Dyy_1)^2 - D = \delta k_0(x_1^2 - Dy_1^2). \tag{4.7.1}$$

If (x_0, y_0) is a solution to the equation $xy_1 - yx_1 = \delta$, then the general solution to this equation is given by

$$x_1 = x_0 + tx \quad \text{and} \quad y_1 = y_0 + ty, \quad t \in \mathbb{Z}.$$

We have

$$xx_1 - Dyy_1 = xx_0 - Dyy_0 + t(x^2 - Dy^2) = xx_0 - Dyy_0 + t\delta k_0.$$

We will choose t such that

$$|xx_1 - Dyy_1| \leq \frac{k_0}{2}. \tag{4.7.2}$$

Denoting by l the positive integer $|xx_1 - Dyy_1|$, from (4.7.1) we obtain

$$x_1^2 - Dy_1^2 = \frac{l^2 - D}{\delta k_0} = \eta h, \tag{4.7.3}$$

where $\eta = \pm 1$ and h is a positive integer.

Using the inequalities $\sqrt{D} < k_0$ and $l < \dfrac{k_0}{2}$, from (4.7.3) it follows that

$$h \leq \frac{\max\{D, l^2\}}{k_0} < \frac{\max\left\{k_0^2, \dfrac{k_0^2}{4}\right\}}{k_0} = \frac{k_0^2}{k_0} = k_0.$$

If $h < k_0$, then we apply again Theorem 3.3.1 and obtain x_1 and y_1 such that $x_1^2 - Dy_1^2 = \eta h$. From the equalities $xy_1 - yx_1 = \delta$ and $xx_1 - Dyy_1 = \pm l$ we deduce the following formulas:

$$x = \frac{-\delta Dy_1 \pm lx_1}{\eta h} \quad \text{and} \quad y = \frac{-\delta x_1 \pm ly_1}{\eta h}. \tag{4.7.4}$$

Hence the integers x and y can be obtained from the equality

$$\eta h(x + y\sqrt{D}) = (x_1 + y_1\sqrt{D})(\pm l - \delta\sqrt{D}).$$

Taking norms in the above equality yields

$$h^2(x^2 - Dy^2) = (x_1^2 - Dy_1^2)(l^2 - D) = \eta h \cdot \eta h \cdot \delta k_0 = h^2 \delta k_0,$$

and so $x^2 - Dy^2 = \delta k_0 = k$.

Therefore, if x and y given in (4.7.4) are integers, then (x, y) is a solution to the general Pell's equation (4.6.1).

If $h > \sqrt{D}$, then we apply again the described procedure.

The considerations above can be summarized in the following algorithm.

Step 1. Find all solutions to the congruence

$$l^2 \equiv D \pmod{k_0},$$

where l is a positive integer and $0 \le l \le \dfrac{k_0}{2}$. Denote by $l_1, l_2, \ldots, l_r$ those satisfying the inequalities $0 \le l \le \dfrac{k_0}{2}$. Set

$$\frac{l_i^2 - D}{\delta k_0} = \eta_i h_i, \quad i = 1, 2, \ldots, r,$$

where $\eta_i = \pm 1$ and h_i is a positive integer.

Step 2. If $k_0 < \sqrt{D}$, apply Theorem 3.3.1.

Step 3. If $k_0 > \sqrt{D}$, consider the equations

$$x_1^2 - Dy_1^2 = \eta_i h_i, \quad i = 1, 2, \ldots, r.$$

From the previous observations we have $0 < h_i < k_0, i =, 1, 2 \ldots, r$.

Step 4. Fix $i \in \{1, 2, \ldots, r\}$.

I. If $h_i < \sqrt{D}$, apply Theorem 3.3.1 to get the solutions to the equation $x_i^2 - Dy_i^2 = \eta_i h_i$. Then the solutions (x, y) are among those given by

$$x = \frac{-\delta D y_i \pm l_i x_i}{\eta_i h_i} \quad \text{and} \quad y = \frac{-\delta x_i \pm l_i y_i}{\eta_i h_i}. \tag{4.7.5}$$

II. If $h_i > \sqrt{D}$, repeat Step 3, replacing δ by η_i and k_0 by h_i. Since $0 < h_i < k_0$, after finitely many operations we will find all solutions to the given equation.

Remark. The two algorithms presented in Sections 4.5 and 4.6 are comparable. None is superior to the other and, moreover, they complete one another. The algorithm in Section 4.5 is preferable for large k's or large D's. The second is more efficient for small k's, for example when k satisfies the inequalities $-\sqrt{D} < k < \sqrt{D}$ (see also Subsection 4.3.4).

4.8   The Diophantine Equation $ax^2 + bxy + cy^2 = N$

The standard approach to solving the equation

$$ax^2 + bxy + cy^2 = N \tag{4.8.1}$$

in relatively prime integers x, y, is via reduction of quadratic forms, as in [127]. There is a parallel approach in [71] which uses continued fractions.

However, in a memoir of 1770, Lagrange, gave a more direct method for solving (4.8.1) when $\gcd(a, b, c) = \gcd(a, N) = 1$ and $D = b^2 - 4ac > 0$ is not a perfect square. This seems to have been largely overlooked. (Admittedly, the necessity part of his proof is long and not easy to follow.)

In [175], equation (4.8.1) is solved when $N = \pm\mu$, where

$$\mu = \min_{(x,y)\neq(0,0)} |ax^2 + bxy + cy^2|.$$

The approach is similar to Lagrange's reduction to the case $N = \pm 1$.

In the doctoral dissertation [157] the equation (4.8.1) is also discussed, using a standard convergent sufficiency condition of Lagrange, which resulted in the restriction $D \geq 16$, thus making rigorous the necessity part of Lagrange's discussion. Only the case $b = 0$ is discussed in detail, along the lines of [57].

The approach using the convergent criterion of Lemma 4.3.1, which results in no restriction on D, while allowing us to deal with the non-convergent case, without having to appeal to the case $\mu = 1$ in [175], whose proof is somewhat complicated.

The continued fractions approach also had the advantage that it produces the solution (x, y) with least positive y from each class, if $\gcd(a, N) = 1$.

The assumption $\gcd(a, N) = 1$ involves no loss of generality. For as pointed out by Gauss in his Disquisitiones (see [95]), there exist relatively prime integers α, γ such that $a\alpha^2 + b\alpha\gamma + c\gamma^2 = A$, where $\gcd(A, N) = 1$. Then, if $\alpha\delta - \beta\gamma = 1$, the unimodular transformation $x = \alpha X + \beta Y$, $y = \gamma X + \delta Y$ converts $ax^2 + bxy + cy^2$ to $AX^2 + BXY + CY^2$. Also, the two forms represent the same integers.

Let us illustrate how we can solve (4.8.1) via the reduction of the quadratic form in the left-hand side. By multiplying both sides of (4.8.1) by $4a$ and completing the square we obtain

$$(2ax + by)^2 - Dy^2 = 4aN, \tag{4.8.2}$$

where $D = b^2 - 4ac$. Assume that $D > 0$ and D is not a perfect square. Then (4.8.2) is a general Pell's equation. Let (u_n, v_n) be the general solution to its Pell's resolvent $u^2 - Dv^2 = 1$ and let (α, β) be the fundamental solution of the class K to the equation $X^2 - DY^2 = 4aN$ (see Section 4.1). Following [39], we have:

Theorem 4.8.1. *All integer solutions $(x_n, y_n)_{n \geq 1}$ to (4.8.1) are given by*

$$\begin{cases} x_n = \dfrac{(\alpha - b\beta)u_n - (b\alpha - D\beta)v_n}{2a} \\ y_n = \beta u_n + \alpha v_n, \end{cases} \tag{4.8.3}$$

where $(u_n, v_n)_{n \geq 1}$ is the solution to the Pell's resolvent, and (α, β) is the fundamental solution of the class K.

Proof. Let $X = 2ax + by$, $Y = y$, and $N_1 = 4aN$. By Theorem 4.1.3, we obtain the general solution to $X^2 - DY^2 = N_1$

$$X_n = \alpha u_n + D\beta v_n \quad \text{and} \quad Y_n = \beta u_n + \alpha v_n.$$

Solving the linear system

$$\begin{cases} 2ax_n + by_n = \alpha u_n + D\beta v_n \\ y_n = \beta u_n + \alpha v_n \end{cases}$$

we get the formulas (4.8.3).

Now let us show that x_n is an integer. To prove this, it is enough to show $2a \mid \alpha - b\beta$ and $2a \mid \alpha b - \beta D$. Indeed, we have $\alpha - b\beta = 2ax$ and

$$\alpha b - \beta D = \alpha b - \beta(b^2 - 4ac) = (\alpha - b\beta)b + 4ac\beta = 2axb + 4ac\beta = 2a(x\beta + 2c\beta),$$

and the properties follow.   □

Example. The equation in Example 5, page 54, in the book [22] is reduced to

$$x^2 - 5xy + y^2 = -3. \tag{4.8.4}$$

In the reference [22] the equation (4.8.4) is solved by Fermat's method of infinite descent. Let us illustrate the method in Theorem 4.8.1 for finding the solutions to (4.8.4). The equation (4.8.4) is equivalent to

$$(2x - 5y)^2 - 21y^2 = -12.$$

Let $X = 2x - 5y$ and $Y = y$. We obtain the general Pell's equation $X^2 - 21Y^2 = -12$. The Pell's resolvent $u^2 - 21v^2 = 1$ has the fundamental solution $(u_1, v_1) = (55, 12)$, hence its general solution $(u_n, v_n)_{n \geq 1}$ is given by $u_n + v_n\sqrt{21} = (55 + 12\sqrt{21})^n$. Using the upper bounds in the Remark after Theorem 4.1.3, we have

$$0 \leq |X| \leq \sqrt{\frac{|N|u_1 + N}{2}} = \sqrt{\frac{12 \cdot 55 - 12}{2}} = \sqrt{6 \cdot 54} = 18,$$

$$0 < Y \leq \sqrt{\frac{|N|u_1 - N}{2D}} = \sqrt{\frac{12 \cdot 55 + 12}{2 \cdot 21}} = \sqrt{16} = 4.$$

Therefore, we obtain the possibilities $|X| = 0, 1, \ldots, 18$ and $Y = 1, 2, 3, 4$. Then we get four solutions $(3, 1), (-3, 1), (18, 4), (-18, 4)$ to the equation $X^2 - 21Y^2 = -12$. It is easy to check that these solutions are not associated with each other and they generates four classes of solutions to the above general Pell's equation. From Theorem 4.8.1 we get all integer solutions to equation (4.8.4):

$$(4u_n + 18v_n, u_n + 3v_n), \quad (u_n + 3v_n, u_n - 3v_n),$$
$$(19u_n + 87v_n, 4u_n + 18v_n), (u_n - 3v_n, 4u_n - 18v_n), n \geq 1.$$

These four classes of solutions give a partition of the solution obtained in the above-mentioned reference [22].

4.9 Thue's Theorem and the Equations $x^2 - Dy^2 = \pm N$

In this section, following the papers [86, 87, 128] and [225] we show how to obtain explicit representations of certain integers in the form $x^2 - Dy^2$ for small $D > 1$, using a constructive version of Thue's theorem based on Euclid's algorithm. Amongst other things, if $u^2 \equiv D \pmod{N}$, $D \not\equiv 1 \pmod{N}$ is solvable and $\gcd(D, N) = 1$, N odd, we show how to find the following representations:

$N = 8k \pm 1$	$N = x^2 - 2y^2$
	$-N = x^2 - 2y^2$
$N = 12k + 1$	$N = x^2 - 3y^2$
$N = 12k - 1$	$-N = x^2 - 3y^2$
$N = 5k + 1$	$N = x^2 - 5y^2$
$N = 5k - 1$	$-N = x^2 - 5y^2$
$N = 24k + 1$ or $24k - 5$	$N = x^2 - 6y^2$
$N = 24k - 1$ or $24k + 5$	$-N = x^2 - 6y^2$
$N = 28k + 1, 28k + 9$ or $28k + 25$	$N = x^2 - 7y^2$
$N = 28k - 1, 28k - 9$ or $28k - 25$	$-N = x^2 - 7y^2$

4.9.1   Euclid's Algorithm and Thue's Theorem

Let a and b be natural numbers, $a > b$, where b does not divide a. Let $r_0 = a$, $r_1 = b$, and for $1 \leq k \leq n$, $r_{k-1} = r_k q_k + r_{k+1}$, where $0 < r_{k+1} < r_k$ and $r_n = 0$. Define sequences $s_0, s_1, \ldots, s_{n+1}$ and $t_0, t_1, \ldots, t_{n+1}$ by

$$s_0 = 1, \ s_1 = 0, \ t_0 = 0, \ t_1 = 1, \ t_{k+1} = -q_k t_k + t_{k-1}, \ s_{k+1} = -q_k s_k + s_{k-1},$$

for $1 \leq k \leq n$. Then the following are easily proved by induction:

  (i) $s_k = (-1)^k |s_k|$, $t_k = (-1)^{k+1} |t_k|$;
 (ii) $0 = |s_1| < |s_2| < \ldots |s_{n+1}|$;
(iii) $1 = |t_1| < |t_2| < \cdots < |t_{n+1}|$;
 (iv) $a = |t_k| r_{k-1} + |t_{k-1}| r_k$ for $1 \leq k \leq n+1$;
  (v) $r_k = s_k a + t_k b$ for $1 \leq k \leq n+1$.

Theorem 4.9.1 (Thue). *Let a and b be integers, $a > b > 1$ with $\gcd(a,b) = 1$. Then the congruence $bx \equiv y \pmod{a}$ has a solution in nonzero integers x and y satisfying $|x| < \sqrt{a}$, $|y| \leq a$.*

Proof. As $r_n = \gcd(a,b) = 1$, $a > \sqrt{a} > 1$, and the remainders $r_0, \ldots, r_n$ in Euclid's algorithm decrease strictly to 1, there is a unique index k such that $r_{k-1} > \sqrt{a} \geq r_k$. Then the equation $a = |t_k| r_{k-1} + |t_{k-1}| r_k$ gives $a \geq |t_k| r_{k-1} > |t_k| \sqrt{a}$. Hence $|t_k| < \sqrt{a}$.

Finally, $r_k = s_k a + t_k b$, so $b t_k \equiv r_k \pmod{a}$ and we can take $x = t_k, y = r_k$. □

4.9.2   The Equation $x^2 - Dy^2 = N$ with Small D

Let $N \geq 1$ be an odd integer, $D > 1$ and not a perfect square. Then a necessary condition for solvability of the equation $x^2 - Dy^2 = \pm N$ with $\gcd(x,y) = 1$ is that the congruence $u^2 \equiv D \pmod{N}$ is solvable. From now on we assume this, together with $\gcd(D,N) = 1$ and $1 < u < N$. Then the Jacobi symbol $\left(\dfrac{D}{N} \right) = 1$.

We note that if N is prime, then $\left(\dfrac{D}{N} \right) = 1$ also implies that $u^2 \equiv D \pmod{N}$ is solvable.

If we take $a = N$ and $b = u$ in Euclid's algorithm, the integers $r_k^2 - Dt_k^2$ decrease strictly for $k = 0, \ldots, n$, from a^2 to $1 - Dt_n^2$ and are always multiples of N. For

$$r_k^2 - Dt_k^2 \equiv t_k^2 u^2 - Dt_k^2 \equiv t_k^2 (u^2 - D) \equiv 0 \pmod{N}.$$

If k is chosen so that $r_{k-1} > \sqrt{N} > r_k$, as in the proof of Thue's theorem, then as

$$N = r_{k-1} |t_k| + r_k |t_{k-1}| > r_{k-1} |t_k|, \tag{4.9.1}$$

we have $|t_k| < \sqrt{N}$ and

$$- DN < r_k^2 - Dt_k^2 < N. \tag{4.9.2}$$

Hence $r_k^2 - Dt_k^2 = -lN$, $-1 < l < D$. In fact, $1 \le l < D$, so

$$- DN < r_k^2 - Dt_k^2 \le -N. \tag{4.9.3}$$

Also $r_k^2 + lN = Dt_k^2$ and hence $Dt_k^2 > lN$,

$$|t_k| > \sqrt{\frac{lN}{D}}. \tag{4.9.4}$$

From equation (4.9.1), $N > r_{k-1}|t_k|$ and inequality (4.9.4) implies

$$r_{k-1} < \sqrt{\frac{DN}{l}}. \tag{4.9.5}$$

4.9.3   The Equations $x^2 - 2y^2 = \pm N$

The assumption $\left(\dfrac{2}{N}\right) = 1$ is equivalent to $N \equiv \pm 1 \pmod 8$. Also $1 \le l < 2$, so $l = 1$ and (4.9.3) gives $r_k^2 - 2t_k^2 = -N$. Hence from equation (4.9.5) with $D = 2$, $r_{k-1} < \sqrt{2N}$ and

$$-N = r_k^2 - 2t_k^2 < r_{k-1}^2 - 2t_{k-1}^2 < r_{k-1}^2 < 2N.$$

Thus $r_{k-1}^2 - 2t_{k-1}^2 = N$.

Example. Let $N = 10000000033$, a prime of the form $8n + 1$. Then $u = 87196273$ gives $k = 10$, $r_{10} = 29015$, $t_{10} = -73627$, $r_9 = 118239$, $t_9 = 44612$ and $r_{10}^2 - 2t_{10}^2 = -N$, $r_9^2 - 2t_9^2 = N$.

Remark. We can express r_{k-1} and t_{k-1} in terms of r_k and t_k. The method is useful later for delineating cases when $D = 5, 6, 7$:

Using the identities

$$(r_k r_{k-1} - Dt_k t_{k-1})^2 - D(t_k r_{k-1} - t_{k-1} r_k)^2 = (r_k^2 - Dt_k^2)(r_{k-1}^2 - Dt_{k-1}^2) \tag{4.9.6}$$

and

$$(-1)^k N = r_k t_{k-1} - r_{k-1} t_k, \tag{4.9.7}$$

we deduce that

$$r_k r_{k-1} - D t_k t_{k-1} = \varepsilon N, \tag{4.9.8}$$

where $\varepsilon = \pm 1$.

From equation (4.9.8) we see that $\varepsilon = 1$, as $t_k t_{k-1} < 0$. Hence

$$r_k r_{k-1} + D T_k T_{k-1} = N, \tag{4.9.9}$$

where $T_k = |t_k|$. Then solving equations (4.9.7) and (4.9.9) with $D = 2$ for r_{k-1} and T_{k-1} yields

$$r_{k-1} = -r_k + 2T_k, \quad T_{k-1} = T_k - r_k.$$

The integers N, $|N| \le 200$, such that the equation $x^2 - 2y^2 = N$ is solvable are:
± 1, ± 2, ± 4, ± 7, ± 8, ± 9, ± 14, ± 16, ± 17, ± 18, ± 23, ± 25, ± 28, ± 31, ± 32, ± 34,
± 36, ± 41, ± 46, ± 47, ± 49, ± 50, ± 56, ± 62, ± 63, ± 64, ± 68, ± 71, ± 72, ± 73,
± 79, ± 81, ± 82, ± 89, ± 92, ± 94, ± 97, ± 98, ± 100, ± 103, ± 112, ± 113, ± 119,
± 121, ± 124, ± 126, ± 127, ± 128, ± 136, ± 137, ± 142, ± 144, ± 146, ± 151, ± 153,
± 158, ± 161, ± 162, ± 164, ± 167, ± 169, ± 175, ± 178, ± 184, ± 188, ± 191, ± 193,
± 194, ± 196, ± 199, ± 200.

The following table presents the numbers $k(2, N)$ of classes of solutions and the sets $\mathcal{K}(2, N)$ of fundamental solutions of classes of $x^2 - 2y^2 = N$, when the equations are solvable and N is positive or negative, $|N| \le 18$ [161].

$x^2 - 2y^2 = N$	$k(2, N)$	$\mathcal{K}(2, N)$
$x^2 - 2y^2 = 2$	1	$(2, 1)$
$x^2 - 2y^2 = -2$	1	$(4, 3)$
$x^2 - 2y^2 = 4$	1	$(6, 4)$
$x^2 - 2y^2 = -4$	1	$(2, 2)$
$x^2 - 2y^2 = 7$	2	$(3, 1)$, $(5, 3)$
$x^2 - 2y^2 = -7$	2	$(1, 2)$, $(5, 4)$
$x^2 - 2y^2 = 8$	1	$(4, 2)$
$x^2 - 2y^2 = -8$	1	$(8, 6)$
$x^2 - 2y^2 = 9$	1	$(9, 6)$
$x^2 - 2y^2 = -9$	1	$(3, 3)$
$x^2 - 2y^2 = 14$	2	$(4, 1)$, $(8, 5)$
$x^2 - 2y^2 = -14$	2	$(2, 3)$, $(6, 5)$
$x^2 - 2y^2 = 16$	1	$(12, 8)$
$x^2 - 2y^2 = -16$	1	$(4, 4)$
$x^2 - 2y^2 = 17$	2	$(5, 2)$, $(7, 4)$
$x^2 - 2y^2 = -17$	2	$(1, 3)$, $(9, 7)$
$x^2 - 2y^2 = 18$	1	$(6, 3)$
$x^2 - 2y^2 = -18$	1	$(12, 9)$

Here are four examples of general Pell's equations $x^2 - 2y^2 = N$, with N big:
$k(2, 833) = 6$ and $\mathcal{K}(2, 833) = \{(29,2), (31,8), (35,14), (49,28), (61,38), (79,52)\}$;
$k(2, 1666) = 5$ and $\mathcal{K}(2, 1666) = \{(42,7), (46,15), (54,25), (62,33), (98,63)\}$;
$k(2, 2737) = 7$ and $\mathcal{K}(2, 2737) = \{(53,6), (55,12), (57,16), (75,38), (107,66),$
$(117,74), (135,88)\}$; $k(2, 3689) = 8$ and $\mathcal{K}(2, 3689) = \{(61,4), (67,20), (71,26),$
$(83,40), (89,46), (109,64), (121,74), (167,110)\}$.

4.9.4 The Equations $x^2 - 3y^2 = \pm N$

The assumption $\left(\dfrac{3}{N}\right) = 1$ is equivalent to $N \equiv \pm 1 \pmod{12}$. From equation
(4.9.3), we have $-3N < r_k^2 - 3t_k^2 \leq -N$. Hence $r_k^2 - 3t_k^2 = -2N$ or $-N$.

Case 1. Assume $N \equiv 1 \pmod{12}$. Then $r_k^2 - 3t_k^2 = -N$ would imply the
contradiction $r_k^2 \equiv -1 \pmod 3$.

Hence $r_k^2 - 3t_k^2 = -2N$ and inequality (4.9.5) implies $r_{k-1} < \sqrt{\dfrac{3N}{2}}$. Hence

$$-2N = r_k^2 - 3t_k^2 < r_{k-1}^2 - 3t_{k-1}^2 < r_{k-1}^2 < \frac{3N}{2}.$$

Consequently, $r_{k-1}^2 - 3t_{k-1}^2 = N$.
We find $2r_{k-1} = -r_k + 3T_k$ and $2T_{k-1} = -r_k + T_k$.

Case 2. Assume $N \equiv -1 \pmod{12}$. Then $r_k^2 - 3t_k^2 = -2N$ would imply the
contradiction $r_k^2 \equiv 0 \pmod 3$. Hence $r_k^2 - 3t_k^2 = -N$ and inequality (4.9.5) implies
$r_{k-1} < \sqrt{3N}$. Hence

$$-N = r_k^2 - 3t_k^2 < r_{k-1}^2 - 3t_{k-1}^2 < r_{k-1}^2 < 3N.$$

Consequently, $r_{k-1}^2 - 3t_{k-1}^2 = N$ or $2N$. However, $r_{k-1}^2 - 3t_{k-1}^2 = N$ implies the
contradiction $r_{k-1}^2 \equiv -1 \pmod 3$. Hence $r_{k-1}^2 - 3t_{k-1}^2 = 2N$.
We find $r_{k-1} = -r_k + 3T_k$ and $T_{k-1} = -r_k + T_k$.
The integers N, $|N| \leq 200$, such that the equation $x^2 - 3y^2 = N$ is solvable are:
1, 4, 6, 9, 13, 16, 22, 24, 25, 33, 36, 37, 46, 49, 52, 54, 61, 64, 69, 73, 78, 81, 88,
94, 96, 97, 100, 109, 117, 118, 121, 132, 141, 142, 144, 148, 150, 157, 166, 169,
177, 181, 184, 193, 196, 198, -2, -3, -8, -11, -12, -18, -23, -26, -27, -32,
-39, -44, -47, -48, -50, -59, -66, -71, -72, -74, -75, -83, -92, -98, -99,
-104, -107, -108, -111, -122, -128, -131, -138, -143, -146, -147, -156,
-162, -167, -176, -179, -183, -188, -191, -192, -194, -200.
The following table contains the numbers $k(3, N)$ of classes of solutions and
the sets $\mathcal{K}(3, N)$ of fundamental solutions of classes of $x^2 - 3y^2 = N$, when the
equations are solvable and N is positive or negative, $|N| \leq 27$ [161].

$x^2 - 3y^2 = N$	$k(3,N)$	$\mathcal{K}(3,N)$
$x^2 - 3y^2 = 4$	1	$(4,2)$
$x^2 - 3y^2 = 6$	1	$(3,1)$
$x^2 - 3y^2 = 9$	1	$(6,3)$
$x^2 - 3y^2 = 13$	2	$(4,1),\ (5,2)$
$x^2 - 3y^2 = 16$	1	$(8,4)$
$x^2 - 3y^2 = 22$	2	$(5,1),\ (7,3)$
$x^2 - 3y^2 = 24$	1	$(6,2)$
$x^2 - 3y^2 = 25$	1	$(10,5)$
$x^2 - 3y^2 = -3$	1	$(3,2)$
$x^2 - 3y^2 = -8$	1	$(2,2)$
$x^2 - 3y^2 = -11$	2	$(1,2),\ (4,3)$
$x^2 - 3y^2 = -12$	1	$(6,4)$
$x^2 - 3y^2 = -18$	1	$(3,3)$
$x^2 - 3y^2 = -23$	2	$(2,3),\ (5,4)$
$x^2 - 3y^2 = -26$	2	$(1,3),\ (7,5)$
$x^2 - 3y^2 = -27$	1	$(9,6)$

Here are five example of equations $x^2 - 3y^2 = N$, with N big: $k(3,121) = 3$ and $\mathcal{K} = \{(13,4),\ (14,5),\ (22,11)\}$; $k(3,253) = 4$ and $\mathcal{K}(3,253) = \{(16,1),\ (19,6),\ (20,7),\ (29,14)\}$; $k(3,1573) = 5$ and $\mathcal{K}(3,1573) = \{(40,3),\ (41,6),\ (44,11),\ (55,22),\ (64,29)\}$; $k(3,3289) = 8$ and $\mathcal{K}(3,3289) = \{(58,5),\ (59,8),\ (61,12),\ (67,20),\ (74,27),\ (86,37),\ (94,43),\ (101,48)\}$' $k(3,3718) = 6$ and $\mathcal{K}(3,3718) = \{(61,1),\ (65,13),\ (71,21),\ (79,29),\ (91,39),\ (119,59)\}$.

4.9.5 The Equations $x^2 - 5y^2 = \pm N$

The assumption $\left(\dfrac{5}{N}\right) = 1$ is equivalent to $N \equiv \pm 1 \pmod 5$. Then from equation (4.9.3), we have $-5N < r_k^2 - 5t_k^2 \le -N$. Hence $r_k^2 - 5t_k^2 = -4N,\ -3N,\ -2N$ or $-N$. We cannot have $r_k^2 - 5t_k^2 = -3N$, as then $\left(\dfrac{5}{3}\right) = 1$. Neither can we have $r_k^2 - 5t_k^2 = -2N$, as N is odd.

Case 1. Assume $N \equiv 1 \pmod 5$. Then $r_k^2 - 5t_k^2 = -N$ would imply the contradiction $r_k^2 \equiv -1 \pmod 5$. Hence $r_k^2 - 5t_k^2 = -4N$. Then r_k and t_k are both odd. Also, inequality (4.9.5) implies $r_{k-1} < \sqrt{\dfrac{5N}{4}}$. Hence $-N \le r_{k-1}^2 - 5t_{k-1}^2 \le N$.

Then as in the remark above, we can show that

(i) if $r_{k-1}^2 - 5t_{k-1}^2 = -N$, then

$$4r_{k-1} = -3r_k + 4T_k, \quad 4T_{k-1} = -r_k + 3T_k,$$

hence $r_k \equiv -T_k \pmod 4$.

(ii) if $r_{k-1}^2 - 5t_{k-1}^2 = N$, then

$$4r_{k-1} = -r_k + 5T_k, \quad 4T_{k-1} = -r_k + T_k,$$

hence $r_k \equiv T_k \pmod 4$.

Case 2. Assume $N \equiv -1 \pmod 5$. Then $r_k^2 - 5t_k^2 = -4N$ would imply the contradiction $r_k^2 \equiv 4 \pmod 5$. Hence $r_k^2 - 5t_k^2 = -N$. Then not both r_k and t_k are odd. Also, inequality (4.9.5) implies $r_{k-1} < \sqrt{5N}$ and we deduce that $-N < r_{k-1}^2 - 5t_{k-1}^2 \le 4N$. Consequently, $r_{k-1}^2 - 5t_{k-1}^2 = N$ or $4N$.

Then, as in the remark above, we can show

(i) if $r_{k-1}^2 - 5t_{k-1}^2 = N$, then

$$r_{k-1} = -2r_k + 5T_k, \quad T_{k-1} = -r_k + 2T_k,$$

hence $r_{k-1} \equiv -2r_k \pmod 5$.

(ii) If $r_{k-1}^2 - 5t_{k-1}^2 = 4N$, then

$$r_{k-1} = -r_k + 5T_k, \quad T_{k-1} = -r_k + T_k,$$

hence $r_{k-1} \equiv -r_k \pmod 5$.

Here is a complete classification of the possible cases:

1. $N = 5k + 1$. Then $r_k^2 - 5t_k^2 = -4N$, while r_k and t_k are odd.

   (i) $r_k \equiv -T_k \pmod 4$. Then $r_{k-1}^2 - 5t_{k-1}^2 = -N$.
   (ii) $r_k \equiv T_k \pmod 4$. Then $r_{k-1}^2 - 5t_{k-1}^2 = N$.

2. $N = 5k - 1$. Then $r_k^2 - 5t_k^2 = -N$, while r_k and t_k are not both odd.

   (i) $r_{k-1} \equiv -2r_k \pmod 5$. Then $r_{k-1}^2 - 5t_{k-1}^2 = N$.
   (ii) $r_{k-1} \equiv -r_k \pmod 5$. Then $r_{k-1}^2 - 5t_{k-1}^2 = 4N$.

The integers N, $|N| \le 200$, such that the equation $x^2 - 5y^2 = N$ is solvable are: $\pm 1, \pm 4, \pm 5, \pm 9, \pm 11, \pm 16, \pm 19, \pm 20, \pm 25, \pm 29, \pm 31, \pm 36, \pm 41, \pm 44, \pm 45,$ $\pm 49, \pm 55, \pm 59, \pm 61, \pm 64, \pm 71, \pm 76, \pm 79, \pm 80, \pm 81, \pm 89, \pm 95, \pm 99, \pm 100,$ $\pm 101, \pm 109, \pm 116, \pm 121, \pm 124, \pm 125, \pm 131, \pm 139, \pm 144, \pm 145, \pm 149, \pm 151,$ $\pm 155, \pm 164, \pm 169, \pm 171, \pm 176, \pm 179, \pm 180, \pm 181, \pm 191, \pm 196, \pm 199.$

The following table gives the numbers $k(5, N)$ and the sets $\mathcal{K}(5, N)$ of the equations $x^2 - 5y^2 = N$, when they are solvable and N is positive or negative, $|N| \le 29$ [161].

$x^2 - 5y^2 = N$	$k(5,N)$	$\mathcal{K}(5,N)$
$x^2 - 5y^2 = 4$	3	$(3,1),\ (7,3),\ (18,8)$
$x^2 - 5y^2 = 5$	1	$(5,2)$
$x^2 - 5y^2 = 9$	1	$(27,12)$
$x^2 - 5y^2 = 11$	2	$(4,1),\ (16,7)$
$x^2 - 5y^2 = 16$	3	$(6,2),\ (14,6),\ (36,16)$
$x^2 - 5y^2 = 19$	2	$(8,3),\ (12,5)$
$x^2 - 5y^2 = 20$	3	$(5,1),\ (10,4),\ (25,11)$
$x^2 - 5y^2 = 25$	1	$(45,20)$
$x^2 - 5y^2 = -4$	3	$(1,1),\ (4,2),\ (11,5)$
$x^2 - 5y^2 = -5$	1	$(20,9)$
$x^2 - 5y^2 = -9$	1	$(6,3)$
$x^2 - 5y^2 = -11$	2	$(3,2),\ (13,6)$
$x^2 - 5y^2 = -16$	3	$(2,2),\ (8,4),\ (22,10)$
$x^2 - 5y^2 = -19$	2	$(1,2),\ (31,14)$
$x^2 - 5y^2 = -20$	3	$(5,3),\ (15,7),\ (40,18)$
$x^2 - 5y^2 = -25$	1	$(10,5)$
$x^2 - 5y^2 = -29$	2	$(4,3),\ (24,11)$

Here are three example of equations $x^2 - 5y^2 = N$, with N big: $k(5,1276) = 11$ and $\mathcal{K}(5,1276) = \{(36,2), (39,7), (41,9), (49,15), (59,21), (76,30), (84,34), (111,47), (141,61), (211,93), (284,126)\}$; $k(5,1936) = 8$ and $\mathcal{K}(5,1936) = \{(46,6), (54,14), (84,32), (116,48), (154,66), (206,90), (294,130), (396,176)\}$; $k(5,9196) = 18$ and $\mathcal{K}(5,9196) = \{(96,2), (99,11), (104,18), (111,25), (121,33), (139,45), (149,51), (176,66), (201,79), (229,93), (264,110), (321,137), (351,151), (429,187), (499,219), (576,254), (671,297), (824,366)\}$.

4.9.6 The Equations $x^2 - 6y^2 = \pm N$

The assumption $\left(\dfrac{6}{N}\right) = 1$ is equivalent to $N \equiv \pm 1 \pmod{24}$ or $N \equiv \pm 5 \pmod{24}$. Then from equation (4.9.3), we have $-6N < r_k^2 - 6t_k^2 \leq -N$. Hence $r_k^2 - 6t_k^2 = -5N,\ -4N,\ -3N,\ -2N$ or $-N$. Only $-4N$ is ruled out immediately and the other possibilities can occur.

As with the case $D = 5$, there is a complete classification of the possible cases:

1. $N = 24k - 1$ or $24k + 5$.

   (i) $r_k \equiv 0 \pmod 3$. Then $r_k^2 - 6t_k^2 = -3N$, $r_{k-1}^2 - 6t_{k-1}^2 = -N$.
   (ii) $r_k \not\equiv 0 \pmod 3$. Then $r_k^2 - 6t_k^2 = -N$.

(a) $r_{k-1} \equiv 0 \pmod 2$. Then $r_{k-1}^2 - 6t_{k-1}^2 = 2N$.
(b) $r_{k-1} \equiv 1 \pmod 2$. Then $r_{k-1}^2 - 6t_{k-1}^2 = 5N$.

2. $N = 24k + 1$ or $24k - 5$:

(i) $r_k \equiv 0 \pmod 2$. Then $r_k^2 - 6t_k^2 = -2N$, $r_{k-1}^2 - 6t_{k-1}^2 = N$.
(ii) $r_k \equiv 1 \pmod 2$. Then $r_k^2 - 6t_k^2 = -5N$.

(a) $r_k \equiv T_k \pmod 5$. Then $r_{k-1}^2 - 6t_{k-1}^2 = N$.
(b) $r_k \equiv -T_k \pmod 5$. Then $r_{k-1}^2 - 6t_{k-1}^2 = -2N$, $r_{k-2}^2 - 6t_{k-2}^2 = N$.

The integers N, $|N| \leq 200$, such that $x^2 - 6y^2 = N$ is solvable are: 1, 3, 4, 9, 10, 12, 16, 19, 25, 27, 30, 36, 40, 43, 46, 48, 49, 57, 58, 64, 67, 73, 75, 76, 81, 90, 94, 97, 100, 106, 108, 115, 120, 121, 129, 138, 139, 142, 144, 145, 147, 160, 163, 169, 171, 172, 174, 184, 190, 192, 193, 196, -2, -5, -6, -8, -15, -18, -20, -23, -24, -29, -32, -38, -45, -47, -50, -53, -54, -60, -69, -71, -72, -80, -86, -87, -92, -95, -96, -98, -101, -114, -116, -125, -128, -134, -135, -141, -146, -149, -150, -152, -159, -162, -167, -173, -180, -188, -191, -194, -197, -200.

The following table gives the numbers $k(6, N)$ and the sets $\mathcal{K}(6, N)$ of equations $x^2 - 6y^2 = N$, when they are solvable and N is positive or negative, $|N| \leq 25$ [161].

$x^2 - 6y^2 = N$	$k(6,N)$	$\mathcal{K}(6,N)$
$x^2 - 6y^2 = 3$	1	$(3, 1)$
$x^2 - 6y^2 = 4$	1	$(10, 4)$
$x^2 - 6y^2 = 9$	1	$(15, 6)$
$x^2 - 6y^2 = 10$	2	$(4, 1)$, $(8, 3)$
$x^2 - 6y^2 = 12$	1	$(6, 2)$
$x^2 - 6y^2 = 16$	1	$(20, 8)$
$x^2 - 6y^2 = 19$	2	$(5, 1)$, $(13, 5)$
$x^2 - 6y^2 = 25$	3	$(7, 2)$, $(11, 4)$, $(25, 10)$
$x^2 - 6y^2 = -2$	1	$(2, 1)$
$x^2 - 6y^2 = -5$	2	$(1, 1)$, $(7, 3)$
$x^2 - 6y^2 = -6$	1	$(12, 5)$
$x^2 - 6y^2 = -8$	1	$(4, 2)$
$x^2 - 6y^2 = -15$	2	$(3, 2)$, $(9, 4)$
$x^2 - 6y^2 = -18$	1	$(6, 3)$
$x^2 - 6y^2 = -20$	2	$(2, 2)$, $(14, 6)$
$x^2 - 6y^2 = -23$	2	$(1, 2)$, $(19, 8)$

Here are three examples of equations $x^2 - 6y^2 = N$, with N big: $k(6, 625) = 5$ and $\mathcal{K}(6, 625) = \{(29,6), (35,10), (55,20), (73,28), (125,50)\}$; $k(6, 2185) = 8$ and $\mathcal{K}(6, 2185) = \{(47,2), (49,6), (61,16), (79,26), (83,28), (113,42), (173,68), (211,84)\}$; $k(6, 9025) = 9$ and $\mathcal{K}(6, 9025) = \{(97,8), (101,14), (133,38), (155,50), (175,60), (209,76), (337,132), (389,154), (475,190)\}$.

4.9.7   The Equations $x^2 - 7y^2 = \pm N$

The assumption $\left(\dfrac{7}{N}\right) = 1$ is equivalent to $N \equiv 1, 3, 9, 19, 25, 27 \pmod{28}$.

As with the case $D = 6$, there is a complete classification of the possible cases:

1. $N = 28k + 1,\ 28k + 9,$ or $28k + 25$.

   (i) $r_k \equiv T_k \pmod 2$. Then $r_k^2 - 7t_k^2 = -6N$.

      (a) $r_k \equiv -T_k \pmod 6$. Then $r_{k-1}^2 - 7t_{k-1}^2 = -3N$.

         (1) $r_{k-1} \equiv -T_{k-1} \pmod 3$. Then $r_{k-2}^2 - 7t_{k-2}^2 = N$.
         (2) $r_{k-1} \equiv T_{k-1} \pmod 3$. Then $r_{k-2}^2 - 7t_{k-2}^2 = 2N$.

      (b) $r_k \equiv T_k \pmod 6$. Then $r_{k-1}^2 - 7t_{k-1}^2 = N$.

   (ii) $r_k \not\equiv T_k \pmod 2$. Then $r_k^2 - 7t_k^2 = -3N$.

      (a) $r_k \equiv -T_k \pmod 3$. Then $r_{k-1}^2 - 7t_{k-1}^2 = N$.
      (b) $r_k \equiv T_k \pmod 3$. Then $r_{k-1}^2 - 7t_{k-1}^2 = 2N$.

2. $N = 28k + 3,\ 28k + 19,$ or $28k + 27$.

   (i) $r_k \equiv T_k \pmod 2$. Then $r_k^2 - 7t_k^2 = -2N$.

      (a) $r_{k-1} \equiv -T_{k-1} \pmod 3$. Then $r_{k-1}^2 - 7t_{k-1}^2 = -N$.
      (b) $r_{k-1} \equiv T_{k-1} \pmod 3$. Then $r_{k-1}^2 - 7t_{k-1}^2 = 3N$.

   (ii) $r_k \not\equiv T_k \pmod 2$. Then $r_k^2 - 7t_k^2 = -N$.

      (a) $r_{k-1} \equiv -T_{k-1} \pmod 3$. Then $r_{k-1}^2 - 7t_{k-1}^2 = 3N$.
      (b) $r_{k-1} \equiv T_{k-1} \pmod 3$. Then $r_{k-1}^2 - 7t_{k-1}^2 = 6N$.

In cases 1(a)(2) and 2(i), the equations $r_{k-2}^2 - 7t_{k-2}^2 = 2N$ and $r_k^2 - 7t_k^2 = -2N$ give rise to equations $x^2 - 7y^2 = N,\ -N$, respectively, if we write $x + y\sqrt{7} = (r_{k-2} + t_{k-2}\sqrt{7})(3 + \sqrt{7})$ and $(r_k + t_k\sqrt{7})/(3 + \sqrt{7})$, respectively. For if $x + y\sqrt{7} = (r + t\sqrt{7})/(3 + \sqrt{7})$, where r and t are odd, then $x = \dfrac{3r - 7t}{2}$ and $y = \dfrac{3t - r}{2}$ are integers and $x^2 - 7y^2 = (r^2 - 7t^2)/2$.

We note that 1(a)(2) cannot occur unless $N \equiv 0 \pmod 3$. Then we have

$$r_{k-1} + \frac{-r_k + 7T_k}{6}, \qquad T_{k-1} = \frac{-r_k + 5T_k}{6} \qquad (4.9.10)$$

$$r_{k-2} = \frac{-r_{k-1} + 7T_{k-1}}{3}, \qquad T_{k-2} = \frac{-r_{k-1} + T_{k-1}}{3}. \qquad (4.9.11)$$

Then (4.9.6) implies $r_{k-1} + T_{k-1} = -r_k + 2T_k \equiv -r_k - T_k \equiv 0 \pmod 3$. Also (4.9.7) implies $r_{k-1} \equiv T_{k-1} \pmod 3$. Hence 3 divides r_{k-1} and T_{k-1} and the equation $r_{k-1}^2 - 7T_{k-1}^2 = -3N$ then implies that 3 divides N.

Example. $N = 57$. The congruence $u^2 \equiv 7 \pmod{57}$ has solutions $u \equiv \pm 8, \pm 11$ (mod 57). Then $u = 0$ gives $k = 2$, $r_1 = 8$, $t_1 = 1$, $r_2 = 1$, $r_2 = 1$, $t_2 = -7$, $r_k^2 - 7t_k^2 = -6N$ and $r_{k-1}^2 - 7t_{k-1}^2 = N$, while $u = 11$ gives $k = 2$, $r_1 = 11$, $t_1 = 1$, $r_2 = 2$, $t_2 = -5$ and $r_k^2 - 7t_k^2 = -3N$ and $r_{k-1}^2 - 7t_{k-1}^2 = 2N$.

The integers N, $|N| \le 200$, such that $x^2 - 7y^2 = N$ is solvable are: 1, 2, 4, 8, 9, 16, 18, 21, 25, 29, 32, 36, 37, 42, 49, 50, 53, 57, 58, 64, 72, 74, 81, 84, 93, 98, 100, 106, 109, 113, 114, 116, 121, 128, 133, 137, 141, 144, 148, 149, 162, 168, 169, 177, 186, 189, 193, 196, 197, 200, -3, -6, -7, -12, -14, -19, -24, -27, -28, -31, -38, -47, -48, -54, -56, -59, -62, -63, -75, -76, -83, -87, -94, -96, -103, -108, -111, -112, -118, -124, -126, -131, -139, -147, -150, -152, -159, -166, -167, -171, -174, -175, -188, -192, -199.

The following table contains the numbers $k(7, N)$ and the sets $\mathcal{K}(7, N)$ of the equations $x^2 - 7y^2 = N$, when they are solvable and N is positive or negative, $|N| \le 29$ [161].

$x^2 - 7y^2 = N$	$k(7,N)$	$\mathcal{K}(7,N)$
$x^2 - 7y^2 = 2$	1	$(3, 1)$
$x^2 - 7y^2 = 4$	1	$(16, 6)$
$x^2 - 7y^2 = 8$	1	$(6, 2)$
$x^2 - 7y^2 = 9$	3	$(4, 1)$, $(11, 4)$, $(24, 9)$
$x^2 - 7y^2 = 16$	1	$(32, 12)$
$x^2 - 7y^2 = 18$	3	$(5, 1)$, $(9, 3)$, $(19, 7)$
$x^2 - 7y^2 = 21$	2	$(7, 2)$, $(14, 5)$
$x^2 - 7y^2 = 25$	1	$(40, 15)$
$x^2 - 7y^2 = 29$	2	$(6, 1)$, $(27, 10)$
$x^2 - 7y^2 = -3$	2	$(2, 1)$, $(5, 2)$
$x^2 - 7y^2 = -6$	2	$(1, 1)$, $(13, 5)$
$x^2 - 7y^2 = -7$	1	$(21, 8)$
$x^2 - 7y^2 = -12$	2	$(4, 2)$, $(10, 4)$
$x^2 - 7y^2 = -14$	1	$(7, 3)$
$x^2 - 7y^2 = -19$	2	$(3, 2)$, $(18, 7)$
$x^2 - 7y^2 = -24$	2	$(2, 2)$, $(26, 10)$
$x^2 - 7y^2 = -27$	4	$(1, 2)$, $(6, 3)$, $(15, 6)$, $(34, 13)$

Here are three examples of equations $x^2 - 7y^2 = N$, with N big: $k(7, 2349) = 10$ and $\mathcal{K}(7, 2349) = \{(51,6), (54,9), (61,14), (82,25), (93,30), (114,39), (131,46), (194,71), (243,90), (282,105)\}$; $k(7, 3249) = 9$ and $\mathcal{K}(7, 3249) = \{(64,11), (71,16), (76,19), (111,36), (132,45), (209,76), (232,85), (281,104), (456,171)\}$; $k(7, 4617) = 12$ and $\mathcal{K}(7, 4617) = \{(68,1), (72,9), (75,12), (93,24), (117,36), (128,41), (163,56), (180,63), (240,87), (348,129), (387,144), (523,196)\}$.

Chapter 5
Equations Reducible to Pell's Type Equations

5.1 The Equations $x^2 - kxy^2 + y^4 = 1$ and $x^2 - kxy^2 + y^4 = 4$

An interesting problem concerning the Pell's equation $u^2 - Dv^2 = 1$ is to study when the second component of a solution (u, v) is a perfect square. This question is equivalent to solving the equation

$$X^2 - DY^4 = 1. \tag{5.1.1}$$

The equation (5.1.1) was intensively studied in a series of papers (see [116–118]). We begin this section by mentioning the main result about the above equation.

Theorem 5.1.1. *For a positive nonsquare integer D there are at most two solutions to the equation (5.1.1). If two solutions exist, and ε_D denotes the fundamental unit in the quadratic field $\mathbb{Q}(\sqrt{D})$, then they are given by (x_1, y_1), (x_2, y_2), $x_1 < x_2$, where $x_1 + y_1^2\sqrt{D} = \varepsilon_D$ and $x_2 + y_2^2\sqrt{D}$ is either ε_D^2 or ε_D^4, with the latter case occurring for only finitely many D.*

Following the recent paper [221] we first prove a generalization of Theorem 5.1.1. We then use this result to completely solve the equations

$$x^2 - kxy^2 + y^4 = 1 \tag{5.1.2}$$

and

$$x^2 - kxy^2 + y^4 = 4. \tag{5.1.3}$$

Let $D = e^2 d$, with e an integer and d a positive squarefree integer. Then $\varepsilon_D = \dfrac{a + b\sqrt{d}}{2}$, where a and b are positive integers with the same parity, and satisfy

© Springer Science+Business Media New York 2015
T. Andreescu, D. Andrica, *Quadratic Diophantine Equations*,
Developments in Mathematics 40, DOI 10.1007/978-0-387-54109-9_5

$a^2 - db^2 = (-1)^\alpha 4$, where $\alpha \in \{0, 1\}$. Define $\lambda_D = \lambda_d$ to be the fundamental solution $u + v\sqrt{d}$ to $X^2 - dY^2 = 1$, with u and v positive integers. Then $\lambda_D = (\varepsilon_D)^c$, where

$$c = \begin{cases} 1 \text{ if } a \text{ and } b \text{ are even and } \alpha = 0 \\ 2 \text{ if } a \text{ and } b \text{ are even and } \alpha = 1 \\ 3 \text{ if } a \text{ and } b \text{ are odd and } \alpha = 0 \\ 6 \text{ if } a \text{ and } b \text{ are odd and } \alpha = 1. \end{cases} \qquad (5.1.4)$$

Lemma 5.1.2 ([55]). *Let D be a nonsquare positive integer. If the equation $X^4 - DY^2 = 1$ is solvable in positive integers X, Y, then either $X^2 + Y\sqrt{D} = \lambda_D$ or λ_D^2. Solutions to $X^4 - DY^2 = 1$ arise from both λ_D and λ_D^2 only for $D \in \{1785, 7140, 28560\}$.*

Lemma 5.1.3. *If there are two solutions to equation (5.1.1), then they are given by $X + Y^2\sqrt{D} = \varepsilon_D, \varepsilon_D^4$ for $D \in \{1785, 28560\}$, and $X + Y^2\sqrt{D} = \varepsilon_D, \varepsilon_D^2$ otherwise.*

Proof. Let $T + U\sqrt{D}$ denote the fundamental solution in positive integers to the Pell's equation $x^2 - Dy^2 = 1$, and for $k \geq 1$ let $T_k + U_k\sqrt{D} = (T + U\sqrt{D})^k$.

If there exist two indices k_1 and k_2 for which U_{k_1} and U_{k_2} are squares, then by Theorem 5.1.1, $(k_1, k_2) = (1, 4)$ or $(k_1, k_2) = (1, 2)$. If there are integers x and y such that $U_1 = x^2$ and $U_4 = y^2$, then since $U_4 = 2T_2U_2$, there exist integers w and z such either $(T_2, U_2) = (w^2, 2z^2)$, or $(T_2, U_2) = (2w^2, z^2)$. The latter case is not possible, since it would imply the existence of three solutions to $X^2 - DY^4 = 1$, contradicting Theorem 5.1.1. In the former case, since $2z^2 = U_2 = 2T_1U_1$, there are integers u and $v > 1$ such that $T_1 = v^2$ and $U_1 = u^2$. We thus have solutions to $X^4 - DY^2 = 1$ arising from both ε_D and ε_D^2. By Lemma 5.1.2, we deduce that $D \in \{1785, 7140, 28560\}$, and since $U_1 = 2$ and $D = 7140$, we have finally that $D \in \{1785, 28560\}$. □

Lemma 5.1.4 ([117]). *The only positive integer solutions to the equation $X^2 - 2Y^4 = -1$ are $(X, Y) = (1, 1), (239, 13)$.*

Lemma 5.1.5 ([45]). *The only positive integer solutions to the equation $3X^4 - 2Y^2 = 1$ are $(X, Y) = (1, 1), (3, 11)$.*

Lemma 5.1.6 ([38]). *With the notations in the proof to Lemma 5.1.3, if $T_k = 2x^2$ for some integer x, then $k = 1$.*

Theorem 5.1.7. *Let D be a nonsquare positive integer with $D \notin \{1785, 7140, 28560\}$. Then there are at most two positive indices k for which $U_k = 2^\delta y^2$ with y an integer and $\delta = 0$ or 1. If two solutions $k_1 < k_2$ exist, then $k_1 = 1$ and $k_2 = 2$, and provided that $D \neq 5$, $T + U\sqrt{D}$ is the fundamental unit in $\mathbb{Q}(\sqrt{D})$, or its square. For $D \in \{1785, 7140, 28560\}$, the only solutions to $U_k = 2^\delta y^2$ are $k = 1$, $k = 2$, and $k = 4$.*

Proof. If one of the equation $x^2 - Dy^4 = 1$, $x^2 - 4Dy^4 = 1$ is not solvable, then the result follows from Lemma 5.1.3 applied to $4D$ and D respectively.

Therefore we may assume that both of these equations are solvable. Let k and l be indices for which $U_k = z^2$ and $U_l = 2w^2$. It follows from the binomial theorem that not both of k and l are odd.

Assume first that k and l are both even. We will show that this leads to $D \in \{1785, 7140, 28560\}$. Letting $l = 2m$, then there are integers $u > 1$ and v such that $T_m = u^2$ and $U_m = v^2$. Then by Lemma 5.1.2, either $m = 1$ or $m = 2$. Also, by Lemma 5.1.3, and the fact that k is even, either $(k, m) = (2, 1)$, $(k, m) = (4, 1)$ and $D \in \{1785, 28560\}$, or else $k = m$. The first case is not possible since it would imply $k = l = 2$, and this contradicts the assumed forms of U_k and U_l. Thus, for $D \notin \{1785, 28560\}$, we have that $k = m$, and furthermore, the only possibility is $k = m = 2$. Since $U_2 = 2T_1 U_1$, there are positive integers a, b for which either $(T_1, U_1) = (a^2, 2b^2)$ or $(T_1, U_1) = (2a^2, b^2)$. From the identity $T_2 = 2T_1^2 - 1$, these two possibilities yield the respective equations $u^2 = 2a^4 - 1$ or $u^2 = 8a^4 - 1$. The equation $u^2 = 8a^4 - 1$ is not solvable modulo 4. By Lemma 5.1.4, the only positive integer solution to the equation $u^2 = 2a^4 - 1$, with $u > 1$, is $u = 239$ and $a = 13$. Therefore, $T_1 = 169$, and $U_1 = 2b^2$ for some integer b. The only choice for b is $b = 1$, which results in $D = 7140$.

We can assume that k and l are of opposite parity. First assume that l is even, $l = 2m$, and that k is odd. Thus, we have that $U_{2m} = 2w^2$. From the identity $U_{2m} = 2T_m U_m$, and the fact that $(T_m, U_m) = 1$, it follows that there are integers u and v such that $T_m = u^2$ and $U_m = v^2$. By Lemma 5.1.2, either $m = 1$ or $m = 2$, and $T_1 + U_1\sqrt{D} = \lambda_D$. Furthermore, by Lemma 5.1.3, either $k = m$ or $k = 1$, $m = 2$. If $k = m$, then since k is odd and $m = 1$ or 2, we have that $k = 1$ and $l = 2$, which is our desired result. On the other hand, if $k = 1$ and $m = 2$, then $l = 4$, and we have that $U_4 = 2w^2$, $U_2 = v^2$, and $T_2 = u^2$. As in the previous paragraph, this leads to $D = 7140$.

Now assume that l is odd and k is even, $k = 2m$. Therefore, $U_{2m} = 2T_m U_m = z^2$, and it follows that there are integers u and v such that either $(T_m, U_m) = (2u^2, v^2)$ or $(T_m, U_m) = (u^2, 2v^2)$. In the first case, Lemma 5.1.3 implies that $(m, k) = (1, 2)$, since U_m and U_k are both squares. Therefore U_1 is a square, and $2^{2\alpha}$ properly divides U_1 for some integer $\alpha \geq 0$, Since $U_l = 2w^2$, $2^{2\beta+1}$ properly divides U_l for some integer $\beta \geq 0$. From the fact that l is odd, the binomial theorem exhibits that the same power of 2 divides U_1 and U_l, thus leading to a contradiction. In the case that $(T_m, U_m) = (u^2, 2v^2)$, Lemma 5.1.2 shows that $m = 1$ or $m = 2$, and that $T_1 + U_1\sqrt{D} = \lambda_D$. Also, by Lemma 5.1.3 applied to $4D$, either $m = l$ or $(l, m) = (1, 2)$. The former possibility leads to $l = 1$ and $k = 2$, which is the desired result. The latter possibility implies that $k = 4$, and that $T_2 = u^2$, $U_2 = 2v^2$, Since $U_2 = 2T_1 U_1$, there are integers a and b such that $T_1 = a^2$, and $U_1 = b^2$. Therefore, $u^2 = T_2 = 2T_1^2 - 1 = 2a^4 - 1$, and by Lemma 5.1.4, it follows that $T_1 = 169$, and hence that $D = 1785$ or $D = 28560$.

It remains to prove that for $D \neq 5$, $T + U\sqrt{D} = T_1 + U_1\sqrt{D}$ is the fundamental unit ε_D in $\mathbb{Q}(\sqrt{D})$, or its square. Letting $T + U\sqrt{D} = \varepsilon_d^c$, then we need to prove that $c = 1$ or $c = 2$, where c is defined in (5.1.4).

Let $D = l^2 d$ with d squarefree. Let $\lambda_d = t + u\sqrt{d}$, and for $k \geq 1$, define $\lambda_d^k = t_k + u_k\sqrt{d}$. Then $T + U\sqrt{D} = \lambda_d^r = t_r + u_r\sqrt{d}$ for some integer r, and

$u_{ir} = lU_i$ for each $i \geq 1$. We assume now that $U_1 = 2^{\delta_1} x^2$ and $U_2 = 2^{\delta_2} y^2$ for some integers x and y. Then $u_r = 2^{\delta_1} l x^2$ and $u_{2r} = 2^{\delta_2} l y^2$. Since $u_{2r} = 2t_r u_r$, it follows that $t_r = z^2$ or $2z^2$ for some integer z. By Lemma 5.1.3 and Lemma 5.1.6, either $r = 1$ or $r = 2$. This implies that c divides 12. We wish to show that 4 does not divide c. If 4 divides c, then $r = 2$ and $N(\varepsilon_d) = -1$, and so there are the integers $V > 1$ and W such that $V^2 - W^2 d = -1$, with $t_2 + u_2 \sqrt{d} = \lambda_d^2 = (V + W\sqrt{d})^4$. Since $r = 2$, Lemma 5.1.6 shows that $t_2 = z^2$. Therefore, $t_2 = z^2 = 8V^4 + 8V^2 + 1$, and as it was shown in [117] that this equation implies $V = 0$, we have a contradiction. Therefore c divides 6, and to complete the proof of the theorem, we need to show that 3 does not divide c.

Assume that 3 divides c. Then $T + U\sqrt{D}$ is the cube of a unit in $\mathbb{Q}(\sqrt{D})$ of the form $\dfrac{a + b\sqrt{D}}{2}$, where a and b are odd, and $a^2 - b^2 D = 4$. Moreover, $T = a\left(\dfrac{a^2 - 3}{2}\right)$ is odd, and so either $T + U\sqrt{D} = X^2 + Y^2\sqrt{D}$ or $T + U\sqrt{D} = X^2 + 2Y^2\sqrt{D}$, i.e., T is not of the form $2X^2$. It follows that $a(a^2 - 3) = 2X^2$. If $(a, a^2 - 3) = 1$, then since a is odd, $a = A^2$ and $a^2 - 3 = 2B^2$ for some integers A, B, which is not possible by considering this last equation modulo 8. Therefore $(a, a^2 - 3) = 3$, and there are integers A, B for which $a = 3A^2$ and $a^2 - 3 = 6B^2$, which results in the equation $3A^4 - 2B^2 = 1$. By Lemma 5.1.5 the only positive integer solutions to this equation are $(A, B) = (1, 1)$ and $(A, B) = (3, 11)$. This shows that either $a = 3$ or $a = 27$. The case $a = 3$ yields $D = 5$, which we have excluded. The case $a = 27$ yields that either $D = 29$ or $D = 725$. It is easily checked that the hypotheses are not satisfied for both of these values of D. □

Corollary 5.1.8. *For $k = 169$, the only positive integer solutions to $x^2 - (k^2 - 1)$ $y^4 = 1$ are $(x, y) = (169, 1), (6525617281, 6214)$.*

For $k > 1$ and $k \neq 169$, the only positive integer solution (x, y) to $x^2 - (k^2 - 1)$ $y^4 = 1$ is $(x, y) = (k, 1)$, unless $k = 2v^2$ for some integer v, in which case $(x, y) = (8v^4 - 1, 2v)$ is the only other solution.

For $k > 1$ there is no positive integer solutions (x, y) to $x^2 - (k^2 - 1)y^4 = 4$, unless $k = v^2$ for some integer v, in which case $(x, y) = (4v^4 - 2, 2v)$ is the only solution.

Proof. The particular case $k = 169$ is easily verified for both equations, and so we assume that $k > 1$ and $k \neq 169$. The fundamental solution to $x^2 - (k^2 - 1)y^2 = 1$ is $(k, 1)$. For $i \geq 1$ define $T_i + U_i\sqrt{k^2 - 1} = (k + \sqrt{k^2 - 1})^i$. There is always the solution $(x, y) = (k, 1)$ to $x^2 - (k^2 - 1)y^4 = 1$, and so by Theorem 5.1.7, if there is another solution, it must come from $T_2 + U_2\sqrt{k^2 - 1} = 2k^2 - 1 + 2k\sqrt{k^2 - 1}$, i.e., $(x, y) = (2k^2 - 1, \sqrt{2k})$. This entails that $2k$ is a perfect square, and hence that $k = 2v^2$ for some integer v, This gives $(x, y) = (8v^4 - 1, 2v)$.

We note that if k is odd, then the minimal solution to $x^2 - \left(\dfrac{k^2 - 1}{4}\right)y^2 = 1$ is $(x, y) = (k, 2)$, from which it follows that for k even or odd, any solution to $x^2 - (k^2 - 1)y^2 = 4$ has both x and y even. Now let (x, y) be a positive integer solution to $x^2 - (k^2 - 1)y^4 = 4$, then x and y are even, and $(u, v) = (x/2, y/2)$ is a positive integer solution to $u^2 - 4(k^2 - 1)v^4 = 1$, and hence there is a positive integer i for which $U_i = 2v^2$. By Theorem 5.1.7, since $U_1 = 1$ is already a square, $i = 2$. Therefore $u + 2v^2\sqrt{k^2 - 1} = T_2 + U_2\sqrt{k^2 - 1} = 2k^2 - 1 + 2k\sqrt{k^2 - 1}$, and hence $k = v^2$. This leads to the solution $(x, y) = (4v^4 - 2, 2v)$ to the equation $x^2 - (k^2 - 1)y^4 = 4$. This completes the proof.   □

Theorem 5.1.9. *Let k be an even positive integer.*

1) *The only solutions to equation (5.1.2) in nonnegative integers (x, y) are $(k, 1)$,*
   $(1, 0)$, $(0, 1)$, unless either k is a perfect square, in which case there are also the
   solutions $(1, \sqrt{k})$, $(k^2 - 1, \sqrt{k})$, or $k = 338$ in which case there are the solutions
   $(x, y) = (114243, 6214)$, $(13051348805, 6214)$.
2) *The only solution in nonnegative integers x, y to the equation (5.1.3) is $(x, y) =$*
   $(2, 0)$, unless $k = 2v^2$ for some integer v, in which case there are also the
   solutions $(2, \sqrt{2k})$, $(2k^2 - 2, \sqrt{2k})$.

Proof. Letting $k = 2s$, then we can rewrite the equation $x^2 - kxy^2 + y^4 = 1$ as

$$(x - sy^2)^2 - (s^2 - 1)y^4 = 1.$$

Aside from the trivial solution $(x, y) = (1, 0)$, Corollary 5.1.8 implies that the only solutions are $y = 1$, $x - sy^2 = \pm s$, unless $s = 2v^2$ for some integer v, in which case there is also the solutions $y = 2v$ and $x - sy^2 = \pm(8v^4 - 1)$, or $k = 338$. In either case, the solutions listed in Corollary 5.1.8 lead to the solutions given in Theorem 5.1.9.

The equation $x^2 - kxy^2 + y^4 = 4$ can be rewritten as

$$(x - sy^2)^2 - (s^2 - 1)y^4 = 4.$$

Corollary 5.1.8 shows that, aside from the trivial solution $(x, y) = (2, 0)$, there is no solution in positive integers unless $s = v^2$ for some integer v, in which case $y = 2v$ and $x - sy^2 = \pm 4v^4 - 2$. It follows that $k = 2v^2$, $y = \sqrt{2k}$, and either $x = 2$ or $x = 2k^2 - 2$.   □

5.2   The Equation $x^{2n} - Dy^2 = 1$

In this section we will discuss the solvability of the equation

$$x^{2n} - Dy^2 = 1, \tag{5.2.1}$$

where D is a nonsquare positive integer and n is an integer greater than 1. When $n = 2$ its solvability was discussed in the papers [51, 231, 232] and in the section above.

In what follows we also employ the equations

$$x^p - 2y^2 = -1,$$ (5.2.2)

and

$$x^p - 2y^2 = 1,$$ (5.2.3)

where p is a prime ≥ 5.

They were studied by elementary methods in the paper [51].

We first present two useful results.

Lemma 5.2.1. *If the equation (5.2.2) has positive integer solution* $(x, y) \neq (1, 1)$, *then* $2p|y$.

Proof. Suppose (x, y) is a positive integer solution of (5.2.2). Then

$$(x + 1) \cdot \frac{x^p + 1}{x + 1} = 2y^2.$$

Since $\left(x + 1, \dfrac{x^p + 1}{x + 1}\right) = 1$ or p, we have

$$x + 1 = 2y_1^2, \quad \frac{x^p + 1}{x + 1} = y_2^2, \quad y = y_1 y_2,$$ (5.2.4)

or

$$x + 1 = 2py_1^2, \quad \frac{x^p + 1}{x + 1} = py_2^2, \quad y = py_1 y_2.$$ (5.2.5)

By the result of [119], $\dfrac{x^p + 1}{x + 1} = y_2^2$, therefore $x = 1$. Thus (5.2.4) gives $x = y = 1$.

For (5.2.5) clearly $p|y$. We will prove $2|y$ with the elementary method given in [51].

If $2 \nmid y$, from (5.2.2), we have $x \equiv 1 \pmod 8$. Put

$$A(t) = \frac{x^p + 1}{x + 1}, \quad t \geq 1 \text{ and } 2 \nmid t,$$

and so $A(t) \equiv 1 \pmod 8$. Let $1 < l < p$ be a positive odd integer. Then there exist an integer r, odd, $0 < r < l$, and $2k$ such that $p = 2kl + r$ or $p = 2kl - r$.

If $p = 2kl + r$, then

$$A(p) = \frac{((x + 1)A(l) - 1)^{2k} x^r + 1}{x + 1} \equiv \frac{x^r + 1}{x + 1} \equiv A(r) \pmod{A(l)},$$ (5.2.6)

since $x^l = (x+1)A(l) - 1$. Now $(A(p), A(l)) = A((p,l)) = A(1) = 1$. Thus (5.2.6) gives

$$\frac{A(p)}{A(l)} = \frac{A(r)}{A(l)}.$$

If $p = 2kl - r$, then $l - r$ is even. Thus

$$\frac{A(p)}{A(l)} = \left(-x^{l-r} \frac{A(r)}{A(l)} \right) = \left(\frac{A(r)}{A(l)} \right),$$

since $A(l) \equiv 1 \pmod 8$ and

$$A(p) = x^{l-r} A(l(2k-1)) + A(l) - x^{l-r} A(r).$$

For l, r, we have

$$\begin{aligned}
l &= 2k_1 r + \varepsilon_1 r_1, & 0 < r_1 < r, \\
r &= 2k_2 r_1 + \varepsilon_2 r_2, & 0 < r_2 < r_1, \\
& \cdots \\
r_{s-1} &= 2k_{s+1} r_s + \varepsilon_{s+1} r_{s+1}, & 0 < r_{s+1} < r_s, \\
r_s &= k_{s+2} r_{s+1},
\end{aligned}$$

where $\varepsilon_i = \pm 1$ $(i = 1, \ldots, s+1)$ and r_i $(i = 1, \ldots, s+1)$ are odd integers. Since $(l, p) = 1$, we have $r_{s+1} = 1$. Hence

$$\left(\frac{A(p)}{A(l)} \right) = \left(\frac{A(r)}{A(l)} \right) = \left(\frac{A(l)}{A(r)} \right) = \left(\frac{A(r_1)}{A(r)} \right) = \left(\frac{A(r)}{A(r_1)} \right) = \left(\frac{A(r_2)}{A(r_1)} \right)$$

$$= \cdots = \left(\frac{A(r_{s+1})}{A(r_s)} \right) = \left(\frac{A(l)}{A(r_s)} \right) = \left(\frac{1}{A(r_s)} \right) = 1.$$

Now, from $\dfrac{x^p + 1}{x + 1} = py_2^2$, we have

$$(py_2)^2 \equiv pA(p) \pmod{A(l)}.$$

Thus

$$\left(\frac{pA(p)}{A(l)} \right) = \left(\frac{p}{A(l)} \right) = \left(\frac{A(l)}{p} \right) = \left(\frac{l}{p} \right) = 1,$$

since $x \equiv -1 \pmod p$ and so $A(l) \equiv l \pmod p$, We have a contradiction if l is taken as an odd quadratic nonresidue of p. This proves the result. $\square$

Lemma 5.2.2. *The equation (5.2.3) has only positive integer solution $x = 3$, $y = 11$ (when $p = 5$).*

Proof. From (5.2.3), we have

$$\frac{x^p - 1}{x - 1} = a^2 \tag{5.2.7}$$

if $p \nmid y$. By the result of [119], the solution of (5.2.7) is $x = 3$ (when $p = 5$). Thus (5.2.3) has positive integer solution $x = 3$, $y = 11$ (when $p = 5$).

If $p | y$, then $2 | y$ by Remark 1. From (5.2.3), $(1 + \sqrt{-2y})(1 - \sqrt{-2y}) = x^p$. With the assumption $(1 + \sqrt{-2y}, 1 - \sqrt{-2y}) = 1$, we have

$$1 + \sqrt{-2y} = (a + b\sqrt{-2})^p, \quad x = a^2 + 2b^2, \tag{5.2.8}$$

where a, b are integers. Since $2 | y$, from (5.2.3), it follows that

$$x \equiv 1 \pmod{8}. \tag{5.2.9}$$

From (5.2.8) and (5.2.9), we have $2 | b$ and $b \neq 0$. Now, (5.2.8) gives

$$1 = a^p + \binom{p}{2} a^{p-2} (b\sqrt{-2})^2 + \cdots + \binom{p}{p-1} a(b\sqrt{-2})^{p-1}. \tag{5.2.10}$$

Thus $a | 1$ and so $a = \pm 1$.

If $a = -1$, then (5.2.10) gives

$$-2 = \binom{p}{2} (b\sqrt{-2})^2 + \cdots + \binom{p}{p-1} (b\sqrt{-2})^{p-1},$$

and so $p | 2$ which is impossible.

If $a = 1$, then we have

$$0 = \binom{p}{2} (b\sqrt{-2})^2 + \cdots + \binom{p}{p-1} (b\sqrt{-2})^{p-1}. \tag{5.2.11}$$

Since $2 | b$ and $b \neq 0$, let $2^{s_k} \| \binom{p}{2k} (b\sqrt{-2})^{2k} \left(1 \leq k \leq \frac{p-1}{2} \right)$, clearly $s_k > s_j$ $(k > j)$. Thus (5.2.11) is impossible. $\qquad\qquad\square$

Theorem 5.2.3. *If $n > 2$ and the negative Pell's equation $u^2 - Dv^2 = -1$ is solvable, then the equation (5.2.1) has only one solution in positive integers: $x = 3$, $y = 22$ (when $n = 5$, $D = 122$).*

Proof. Let $\Omega = u_0 + v_0\sqrt{D}$ be the smallest solution to the equation $u^2 - Dv^2 = -1$, $\overline{\Omega} = u_0 - v_0\sqrt{D}$, and let $\eta = U_0 + V_0\sqrt{D}$ be the fundamental solution of the equation $U^2 - DV^2 = 1$, $\overline{\eta} = U_0 - V_0\sqrt{D}$. Then, we have $\eta = \Omega^2$.

Suppose (x, y) is any positive integer solution of (5.2.1). Then

$$x^n = \frac{\eta^m + \overline{\eta}_m}{2} = \frac{\Omega^{2m} + \overline{\Omega}^{2m}}{2}, \quad m > 0. \qquad (5.2.12)$$

Clearly, without loss of generality, we may assume that $n = 4$ or $n = p$ (p is odd prime).

(a) If $n = 4$, then (5.2.12) gives

$$x^4 = 2 \left(\frac{\Omega^m + \overline{\Omega}^m}{2} \right)^2 - (-1)^m,$$

and so $x = 1$, $m = 0$, which is impossible since $m > 0$.

(b) If $n = p$ (p is odd prime), then (5.2.12) gives

$$x^p = 2 \left(\frac{\Omega^m + \overline{\Omega}^m}{2} \right)^2 - (-1)^m. \qquad (5.2.13)$$

(b.1) When $2 \mid m$, let $m = 2s$, $s > 0$; then (5.2.13) gives

$$x^p + 1 = 2 \left(\frac{\Omega^{2s} + \overline{\Omega}^{2s}}{2} \right)^2. \qquad (5.2.14)$$

Suppose $p = 3$. Then by (5.2.14), we have (see [218])

$$x = \frac{\Omega^{2s} + \overline{\Omega}^{2s}}{2} = 1, \qquad (5.2.15)$$

and

$$x = 23, \quad \frac{\Omega^{2s} + \overline{\Omega}^{2s}}{2} = 78. \qquad (5.2.16)$$

Clearly (5.2.15) is impossible, since $s > 0$, and (5.2.16) is also impossible since

$$\frac{\Omega^{2s} + \overline{\Omega}^{2s}}{2} = 2 \left(\frac{\Omega^s + \overline{\Omega}^s}{2} \right)^2 - (-1)^s$$

is odd.

Thus $p > 3$. For (5.2.14) we have $2p|\dfrac{\Omega^{2s} + \overline{\Omega}^{2s}}{2}$ by Lemma 5.2.1.

However, $2|\dfrac{\Omega^{2s}+\overline{\Omega}^{2s}}{2}$ is impossible.

(b.2) When $2 \nmid m$, we have

$$x^p - 1 = 2\left(\frac{\Omega^m + \overline{\Omega}^m}{2}\right)^2, \tag{5.2.17}$$

and so $x = 1$, $(\Omega^m + \overline{\Omega}^m)/2 = 0$ when $p = 3$ (see [218]). If $p > 3$, then (5.2.17) gives $x = 3$, $(\Omega^m + \overline{\Omega}^m)/2 = 11$ (when $p = 5$) by Lemma 5.2.2. Thus (5.2.1) has only positive integer solution $x = 3$, $y = 22$ (when $n = 5, D = 122$).   □

Theorem 5.2.4. *If $\eta = U_1 + V_1\sqrt{D}$ is the fundamental solution to Pell's equation $U^2 - DV^2 = 1$, then the positive integer solutions to equation (5.2.1) do not satisfy*

$$x^n + y\sqrt{D} = \eta^{4m}, \quad n > 2, \ m > 0.$$

Proof. If

$$x^n + y\sqrt{D} = \eta^{4m}, \quad n > 2, m > 0,$$

then we have

$$x^n = \frac{\eta^{4m} + \overline{\eta}^{4m}}{2} = 2\left(\frac{\eta^{2m} + \overline{\eta}^{2m}}{2}\right)^2 - 1. \tag{5.2.18}$$

By Lemma 5.2.1, the equality (5.2.18) is impossible since $2 \nmid (\eta^{2m} + \overline{\eta}^{2m})/2$ and $m > 0$.   □

As applications of the above results we will discuss now some interesting problems in number theory.

In 1939 (see [70]) it was conjectured that the equation

$$\binom{n}{m} = y^k, \quad n > m \geq 2, \ k \geq 3 \tag{5.2.19}$$

has no integer solution. In [70] it is proved that the conjecture is right when $m > 4$, leaving the cases $m = 2$ and $m = 3$ unsolved. Now, we can deduce the following result:

Corollary 5.2.5. *The equation*

$$\binom{n}{2} = y^{2k}$$

has no positive integer solution (n, y) with $n \geq 3$ and $k \geq 2$.

Proof. From $\binom{n}{2} = \dfrac{n(n-1)}{2} = y^{2k}$, we have

$$n - 1 = 2y_1^{2k}, \quad n = y_2^{2k}, \quad y = y_1 y_2,$$

or

$$n - 1 = y_2^{2k}, \quad n = 2y_1^{2k}, \quad y = y_1 y_2.$$

Hence

$$y_2^{2k} \mp 1 = 2y_1^{2k}. \tag{5.2.20}$$

If $2|k$, then (5.2.20) clearly gives $|y_1 y_2| \leq 1$; on the other hand, $n \geq 3$ and $\binom{n}{2} = y^{2k}$ imply $|y| = |y_1 y_2| > 1$. Here we have a contradiction. If $2 \nmid k$, $k \geq 2$, we may conclude from Theorem 5.2.3 and Lemma 5.2.1 that (5.2.20) is impossible. $\square$

Define the generalized Pell sequence by

$$x_0 = 1, \quad x_1 = a, \quad x_{n+2} = 2ax_{n+1} - x_n, \tag{5.2.21}$$

where a is an integer greater than 1.

Corollary 5.2.6. *The equation*

$$x_{4n} = y^m$$

has no positive integer solution (n, y), *when* $m \geq 3$.

Proof. From (5.2.21) we have $x_n = \dfrac{\alpha^n + \overline{\alpha}^n}{2}$, $n \geq 0$, where $\alpha = a + \sqrt{a^2 - 1}$ and $\overline{\alpha} = a - \sqrt{a^2 - 1}$ are roots of the trinomial $z^2 - 2az + 1$. Let $a^2 - 1 = Db^2$, where $D > 0$ is squarefree and b is positive integer. Then

$$\alpha = a + b\sqrt{D}, \quad \overline{\alpha} = a - b\sqrt{D} \quad \text{and} \quad \alpha\overline{\alpha} = 1.$$

Thus $y_n = \dfrac{\alpha^n - \overline{\alpha}^n}{2\sqrt{D}}$ satisfies

$$x_n^2 - Dy_n^2 = 1. \tag{5.2.22}$$

By Theorem 5.2.4, the relation (5.2.22) is impossible when $4|n$, $x_n = y^m$ and $m \geq 3$. $\square$

Clearly, if $a = 2u^2 + 1$ $(u > 0)$, then $Db^2 = a^2 - 1 = 4u^2(u^2 + 1)$. Thus $2u|b$. Letting $b = 2uv$, we have $u^2 + 1 = Dv^2$. Hence, using Theorem 5.2.3, we obtain

Corollary 5.2.7. *For the generalized Pell sequence*

$$x_0 = 1, \quad x_1 = 2u^2 + 1, \quad x_{n+2} = 2(2u^2 + 1)x_{n+1} - x_n,$$

where u is positive integer, x_n is never an mth power if $m \geq 3$, except for $x_1 = 2 \cdot 11^2 + 1 = 3^5$.

Remarks. 1) In the paper [129] it is studied the equation (5.2.1), where $n = p$ is a prime. The main two results given there are:

1. If $p = 2$ and $D > \exp(64)$, then (5.2.1) has at most one positive integer solution (x, y).
2. If $p > 2$ and $D > \exp(\exp(\exp(\exp(10))))$, then $2 \nmid m$, where (x, y) is a solution to (5.2.1) expressed as

$$x^p + y\sqrt{D} = \varepsilon_1^m$$

and $\varepsilon_1 = u_1 + v_1\sqrt{D}$ is the fundamental solution to the Pell's equation $u^2 - Dv^2 = 1$.

2) In the paper [189] it is studied the equation $m^4 - n^4 = py^2$, where $p \geq 3$ is a prime, and then the equations $x^4 + 6px^2y^2 + p^2y^4 = z^2$, $c_k(x^4 + 6px^2y^2 + p^2y^4) + 4pd_k(x^3y + pxy^3) = z^2$, for $p \in \{3, 7, 11, 19\}$ and (c_k, d_k) is a solution to the Pell's equation $c^2 - pd^2 = 1$ or to the negative Pell's equation $c^2 - pd^2 = -1$.
3) In the paper [190] is considered the equation $x^4 - q^4 = py^3$, with the following conditions: p and q are distinct primes, x is not divisible by $p, p \equiv 11 \pmod{12}$, $q \equiv 1 \pmod 3$, x is not divisible by p, $p \equiv 11 \pmod{12}$, $q \equiv 1 \pmod 3$, $\bar{p}$ is a generator of the group $(\mathbb{Z}_q^*, \cdot)$, and 2 is a cubic residue mod q. This equation has been solved in the general case in the paper [121].

5.3 The Equation $x^2 + (x + 1)^2 + \cdots + (x + n - 1)^2$ $= y^2 + (y + 1)^2 + \cdots + (y + n + k - 1)^2$

In the paper [2] the relation $5^2 = 3^2 + 4^2$ was considered as the simplest solution in positive integers to various Diophantine equations, in particular, as the simplest solution for the case $n = 1$ to

$$x^2 + (x + 1)^2 + (x + 2)^2 + \cdots + (x + n - 1)^2$$
$$= y^2 + (y + 1)^2 + (y + 2)^2 + \cdots + (y + n)^2, \quad (5.3.1)$$

i.e., the case where the sum of n consecutive squares equals the sum of $n+1$ consecutive squares. The complete set of solutions of (5.3.1) for all positive integers n, for which n and $n+1$ are squarefree, was given in [2] and [4].

The relation $5^2 = 3^2 + 4^2$ may also be considered as the simplest solution in positive integers for the case $k = 2$ of the sum of k consecutive squares is a perfect square. This problem is treated in [5].

In this section, we consider the equation (5.3.1) as the special case for $k = 1$ of

$$x^2 + (x+1)^2 + (x+2)^2 + \cdots + (x+n-1)^2$$
$$= y^2 + (y+1)^2 + (y+2)^2 + \cdots + (y+n+k-1)^2, \qquad (5.3.2)$$

i.e., the case where the sum of n consecutive squares equals the sum of $n+k$ consecutive squares, and present results for $k \geq 2$. We will use the approach in [3].

Theorem 5.3.1. *The equation (5.3.2) is not solvable for $k \equiv 3, 4,$ or 5 (mod 8).*

Proof. The sum S of squares of n consecutive integers, modulo 4, is listed in the table:

n	1	2	3	4	5	6	7	8
S (mod 4)	0 or 1	1	1 or 2	2	2 or 3	3	0 or 3	0

Clearly, beginning with $n = 9$, the row for S (mod 4) must repeat itself and continue to do so with the length of the period equal to 8. Now, if the sum of n consecutive squares is to equal the sum of $n+3$ consecutive squares, there must be, for some n, a number in the S-row which also appears in the S-row for $n+3$. This, however, is not the case for any value of n. Since the column of entries in the S-row repeats with period 8, the same is true for any value of $k \equiv 3$ (mod 8). The same argument can be used to prove the nonexistence of solutions for $k \equiv 4$ or 5 (mod 8). $\square$

Theorem 5.3.2. *The equation (5.3.2) is not solvable for $k \equiv 7, 11, 16,$ or 20 (mod 27).*

Proof. Using the formula for the sum of the first n squares, (5.3.2) can be rewritten as

$$nx^2 + n(n-1)x + n(n-1)(2n-1)/6$$
$$= (n+k)y^2 + (n+k)(n+k-1)y + (n+k)(n+k-1)(2n+2k-1)/6$$

or

$$n(2x+n-1)^2 = (n+k)(2y+n+k-1)^2 + kn^2 + k^2n + k(k^2-1)/3. \qquad (5.3.3)$$

Letting

$$z = 2x + n - 1 \quad \text{and} \quad w = 2y + n + k - 1,$$

we can rewrite (5.3.3) as

$$nz^2 = (n + k)w^2 + kn^2 + k^2n + k(k^2 - 1)/3. \tag{5.3.4}$$

Considering first the case where $k \equiv$ (mod 27), we substitute into (5.3.4) $k = 27\lambda + 7$ and obtain

$$nz^2 = (n + 27\lambda + 7)w^2 + (27\lambda + 7)n^2 + (27\lambda + 7)^2 n$$

$$+ (27\lambda + 7)(243\lambda^2 + 126\lambda + 16). \tag{5.3.5}$$

If $n \equiv 0$ (mod 3), the left-hand side is congruent to 0, modulo 3, while the right-hand side is congruent to $w^2 + 1$, modulo 3, so that $w^2 \equiv 2$, which is a contradiction. If $n \equiv 2$ (mod 3), a contradiction is similarly obtained as the left-hand side is congruent to $2z^2$, modulo 3, and the right-hand side congruent to 1, modulo 3.

If in (5.3.5), $n \equiv 1$ (mod 3), we obtain

$$z^2 \equiv 2w^2 \pmod{3},$$

which is satisfies only if $z \equiv w \equiv 0$ (mod 3), so that we can set

$$z = 3z' \quad \text{and} \quad w = 3w', \quad n = 3m + 1,$$

which, when substituted into (5.3.5) yields

$$(3m + 1)9z'^2 = (3m+27\lambda+8)9w'^2+(27\lambda+7)(3m+1)^2+(27\lambda+7)^2(3m+1)$$

$$+(27\lambda + 7)(243\lambda^2 + 126\lambda + 16),$$

which immediately leads to a contradiction, since the left-hand side is congruent to 0, modulo 9, while the right-hand side is congruent to 6, modulo 9.

By substituting into (5.3.4) $k = 27\lambda + 11$, $k = 27\lambda + 16$, and $k = 27\lambda + 20$, and using the procedure shown above for $k = 27\lambda + 7$, we can similarly show that there are no solutions for (5.3.2) if $k \equiv 11, 16$, or 20 (mod 27). □

We now turn to the question of finding values of k for which solutions to (5.3.2) exist. Such solutions can be obtained either by an analysis of (5.3.4) which is equivalent to (5.3.2) or by programming a computer to find solutions directly from (5.3.2). Using both methods, all values of $k \leq 100$, not excluded by Theorems 5.3.1 and 5.3.2, were considered and solutions found for the values of k indicated in the following table. In each case, we also list a solution for the indicated value of n and give x and y, as defined by (5.3.2).

k	n	x	y	k	n	x	y	k	n	x	y
6	5	28	15	39	2	25169	5539	64	16	740	294
8	3	137	67	40	5	378	104	71	2	378	23
9	3	23	6	41	11	1551	690	72	10	163	13
10	5	25	8	42	25	77	18	73	73	217	102
15	2	2743	933	46	1	3854	539	78	5	754	143
17	17	33	11	48	2	2603	496	79	79	312	166
18	3	127	38	49	2	210	14	80	3	2196	376
22	11	38	7	50	3	243	30	81	3	1257	195
23	2	8453	2379	54	39	160	67	86	43	188	51
24	2	24346	6740	55	55	128	51	87	2	510565	76493
25	25	123	70	56	14	151	33	89	89	227	97
26	3	1417	442	57	19	183	56	90	3	3521	586
31	4	196	49	58	11	36927	14712	94	33	608	253
32	3	239723	70167	62	25	5316	2813	95	2	716	51
33	11	313	137	63	2	236	5	96	1	679	15

Thus the first entry in the table means that

$$28^2 + 29^2 + 30^2 + 31^2 + 32^2 = 15^2 + 16^2 + \cdots + 25^2.$$

In the table above, no attempt was made to list for each given value of k the smallest value of n for which there exists a solution, since, as is evident from an inspection of the table, small values of n are frequently associated with very large values of x and y.

For each pair of values (k, n) for which a solution is given in the above table, infinitely many additional solutions can be obtained as follows.

Letting

$$kn^2 + k^2n + \frac{k(k^2 - 1)}{3} = A$$

equation (5.3.4) can be rewritten as $nz^2 - (n + k)w^2 = A$ or, multiplying both sides by n,

$$u^2 - Dw^2 = N, \tag{5.3.6}$$

where $u = nz$, $D = n(n + k)$, $N = nA$.

Now if (h_m, k_m) is the general solution of Pell's equation

$$h^2 - Dk^2 = 1 \tag{5.3.7}$$

and (u_1, w_1) is any solution to the general Pell's equation (5.3.6), then

$$u_m = u_1 h_m \pm D w_1 k_m, \quad w_m = u_1 k_m \pm w_1 h_m, \quad m \geq 0 \qquad (5.3.8)$$

are solutions to the equation (5.3.6). For details we refer to Section 4.3.

On the other hand, equations (5.3.8) do not necessarily give all solutions for a given pair of values (k, n). Indeed, any attempt to find all solutions for given n and k is bound to lead to presently unsurmountable difficulties as complete solutions of (5.3.6) are available only for $n < \sqrt{D}$, and this condition will generally not be satisfied for $k > 1$ (see Section 4.3.4).

While in the case $k = 1$ (see [4] and [5]) it was shown that solutions exist for all values of n, for which n and $n + 1$ are squarefree, it can easily be shown that for $k > 1$, even if there exists a solution for some n, there may be none for others. Thus, for example, it can be shown that for $k = 6$, solutions can exist only if $n \equiv 1$ or 5 (mod 6). Such facts can be established by arguments similar to those used in the proof of Theorem 5.3.1, making use of the facts that the sequence of values of S (mod 4) in the table of that proof has period 8 and that, if a similar table were constructed for the sequence of values of S (mod 3), it would have period 9.

It is of interest to note that Theorems 5.3.1 and 5.3.2, together with the table of solutions of equation (5.3.2) for $k \leq 100$, presented above, answers for all but 6 values of k the question as to whether or not a solution of (5.3.2) exists for values of $k \leq 100$. These 6 values are $k = 2, 14, 30, 66, 82, 98$. No general method for proving the existence or nonexistence of solutions in individual cases seems to suggest itself. To illustrate typical proofs, we show below the ones for the case $k = 2$, where the knowledge of the Jacobi symbol is involved leads to a solution, and $k = 14$, where an analysis of the highest power of 2 dividing the constant term of (5.3.4) solves the problem.

Theorem 5.3.3. *The equation (5.3.2) is not solvable for* $k = 2$.

Proof. The sum S of the squares of n consecutive integers, modulo 12, is listed in the following table.

n	S (mod 12)
1	$0, 1, 4,$ or 9
2	1 or 5
3	2 or 5
4	2 or 6
5	$3, 6, 7,$ or 10
6	7
7	$4, 7, 8,$ or 11
8	0 or 8
9	0 or 9
10	1 or 9
11	$1, 2, 5,$ or 10
12	2

Obviously, for $13 \leq n \leq 24$, the values of S (mod 12) are those of the above table increased by 2, while those for $25 \leq n \leq 36$ are those of the above table increased by 4, etc. From this, it is immediately seen that all values of n except $n \equiv 5$ or 11 (mod 12) cannot give solutions.

Now substituting $n = 12m + 5$ into (5.3.4) yields

$$(12m + 5)z^2 = (12m + 7)w^2 + 72(2m + 1)^2,$$

so that

$$-2z^2 \equiv 72(2m + 1)^2 \ (\text{mod } 12m + 7)$$

or

$$z^2 \equiv -36(2m + 1)^2 \ (\text{mod } 12m + 7),$$

which means that the Jacobi symbol $(-1/12m + 7)$ must have the value $+1$, which is a contradiction.

An entirely similar analysis for the case $n = 12m + 11$ leads to another contradiction.   □

Theorem 5.3.4. *The equation (5.3.2) is not solvable for $k = 14$.*

Proof. By simple congruence analysis we find that all values of n except $n \equiv 1$ (mod 4) can be excluded. Now substituting $n = 4m + 1$ into (5.3.4) yields

$$(4m + 1)z^2 = (4m + 15)w^2 + 14(4m + 1)^2 + 196(4m + 1) + 910. \qquad (5.3.9)$$

Considering the above equation, modulo 4, we obtain $z^2 \equiv 3w^2$ (mod 4), which shows that z and w must both be even.

Now, if m is even, then (5.3.9) can be rewritten by letting $m = 2m'$ as

$$(8m' + 1)z^2 = (8m' + 15)w^2 + 224(4m'^2 + 8m' + 5).$$

Since letting $z = 2z'$, $w = 2w'$ leads exactly as shown above to the conclusion that z' and w' must be even, we let $z = 4z'$ and $w = 4w'$ and divide by 16 to obtain

$$(8m' + 1)z'^2 = (8m' + 15)w'^2 + 14(4m'^2 + 8m' + 5).$$

Considering this equation modulo 8, we obtain

$$z'^2 = -w'^2 + 6 \ (\text{mod } 8),$$

which is a contradiction, since the left-hand side is congruent to 0,1, or 4 (mod 8), while the right-hand side is congruent to 2,5, or 6 (mod 8).

Now m is odd, then (5.3.9) can be rewritten by letting $m = 2m' + 1$ as

$$(8m' + 5)z'^2 = (8m' + 19)w'^2 + 448(2m'^2 + 6m' + 5)$$

or, diving both sides by 64 and letting $w = 8w'$, $z = 8z'$, as

$$(8m' + 5)z'^2 = (8m' + 19)w'^2 + 7(2m'^2 + 6m' + 5).$$

Considering this equation modulo 4, we obtain

$$z'^2 \equiv 3w'^2 + 3 \pmod 4,$$

which again is a contradiction.   □

5.4   The Equation $x^2 + 2(x+1)^2 + \cdots + n(x+n-1)^2 = y^2$

In this section, following [229], we will discuss the equation

$$x^2 + 2(x+1)^2 + \cdots + n(x+n-1)^2 = y^2 \qquad (5.4.1)$$

determining the values of n for which it has finitely or infinitely many positive integer solutions (x, y).

Theorem 5.4.1. *For each $n \geq 2$ the equation (5.4.1) is solvable and it has infinitely many solutions unless $\dfrac{n(n+1)}{2}$ is a perfect square.*

Proof. The equation (5.4.1) can be written immediately into the form

$$\frac{n(n+1)}{2}x^2 + \frac{2(n-1)n(n+1)}{3}x + \frac{(n-1)n(n+1)(3n-2)}{12} = y^2. \qquad (5.4.2)$$

The substitutions

$$k = \frac{n(n+1)}{2}, \quad u = 3x + 2(n-1), \quad v = \frac{3y}{k}$$

along with the observation

$$\frac{(n-1)n(n+1)(n+2)}{4} = k(k-1)$$

reduce (5.4.2) to the general Pell's equation

$$u^2 - kv^2 = 1 - k. \qquad (5.4.3)$$

For all positive integral values of k, the equation (5.4.3) admits the solution $u_0 = 2n+1$, $v_0 = 3$, corresponding to the solution $x = 1$, $y = k$ of (5.4.1) which is the familiar formula for the sum of the first n cubes. Thus (5.4.1) has always at least one solution.

Now, let $k = \dfrac{n(n+1)}{2}$ be a nonsquare. In this case the Pell's equation

$$U^2 - kV^2 = 1 \qquad\qquad (5.4.4)$$

has the solutions $(U_m, V_m)_{m \geq 0}$ given in Sections 3.2, 3.3, 3.4.

By using the theory of general Pell's equation developed in Chapter 4, it follows that if (u_0, v_0) is a solution of (5.4.3), then

$$u_m = u_0 U_m + k v_0 V_m, \qquad v_m = v_0 U_m + u_0 V_m, \qquad m = 0, 1, \ldots \qquad (5.4.5)$$

are solutions to (5.4.3).

These will give solutions to (5.4.1) in all cases where

$$x = \frac{u_m + 2 - 2n}{3} \quad \text{and} \quad y = \frac{k v_m}{3} \qquad\qquad (5.4.6)$$

are integers. We proceed to examine these.

If $n \equiv 1 \pmod 3$, then $k \equiv 1$, $u_0 \equiv 0$, $v_0 \equiv 0 \pmod 3$, and each u_m, v_m given in (5.4.5) will satisfy $u_m \equiv 0$, $v_m \equiv 0 \pmod 3$, which imply that x and y in (5.4.6) are integers.

If $n \equiv 2 \pmod 3$, then $k \equiv 0$, $u_0 \equiv 2$, $v_0 \equiv 0$, $U_0^2 \equiv 1 \pmod 3$ hence $U_0 \equiv 1$ or $2 \pmod 3$. For $U_0 \equiv 1 \pmod 3$ the relations (5.4.5) show that $u_m \equiv 2$, $u_m + 2 - 2n \equiv 0$, $k v_m \equiv 0 \pmod 3$, hence x and y in (5.4.6) are integers. For $U_0 \equiv 2 \pmod 3$ we have $u_m \equiv 1 \pmod 3$, and x is not an integer. However, in this case, from (5.4.5), $u_{m+1} \equiv 2 \pmod 3$ so that the corresponding x and y are integers.

Analogous study of the case $n \equiv 0 \pmod 3$ gives a similar result. Hence, in all cases, at least alternate members of the infinite sequence of solutions to (5.4.3) give integral values of x, y which satisfy the equation (5.4.1). $\qquad\square$

Remark. One may determine explicitly (see [15]) the integers n for which $k = \dfrac{n(n+1)}{2}$ is a perfect square. This reduces to finding the solutions to the equations $n(n+1) = 2s^2$ or, equivalently, $(2n+1)^2 - 8s^2 = 1$. The last Pell's equation has solutions $(2n_l + 1, s_l)_{l \geq 1}$, where

$$2n_l + 1 = \frac{1}{2}\left[\left(3 + \sqrt{8}\right)^l + \left(3 - \sqrt{8}\right)^l\right]. \qquad\qquad (5.4.7)$$

From (5.4.7) it follows that all positive integers n with the above property are given by

$$
n_l = \begin{cases} \left[\dfrac{\left(\sqrt{2}+1\right)^l - \left(\sqrt{2}-1\right)^l}{2} \right]^2 & \text{if } l \text{ is odd} \\[6mm] 2 \left[\dfrac{\left(1+\sqrt{2}\right)^l - \left(1-\sqrt{2}\right)^l}{2\sqrt{2}} \right]^2 & \text{if } l \text{ is even} \end{cases} \tag{5.4.8}
$$

5.5   The Equation $(x^2 + a)(y^2 + b) = F^2(x,y,z)$

In this section we study the general class of Diophantine equations

$$
(x^2 + a)(y^2 + b) = F^2(x, y, z) \tag{5.5.1}
$$

where $F : \mathbb{Z}_+ \times \mathbb{Z}_+ \times \mathbb{Z} \to \mathbb{Z}^*$ is a given function and a, b are nonzero integers satisfying $|a|, |b| \le 4$.

It is clear that if only one of $x^2 + a$ or $y^2 + b$ is a perfect square, then the equation (5.5.1) is not solvable. In the given hypothesis, $x^2 + a$ and $y^2 + b$ are simultaneously nonzero perfect squares only if $|a| = 3$ and $|b| = 3$ in which situation (x, y) is one of the pairs $(1,1), (1,2), (2,1), (2,2)$. For these pairs we must have

$$
F(1,1,z) = \pm 4, \quad F(1,2,z) = \pm 2, \quad F(2,1,z) = \pm 2, \quad F(2,2,z) = \pm 1. \tag{5.5.2}
$$

It remains to find z from the corresponding equations in (5.5.2), a problem that is strictly dependent upon the function F.

In order to have a unitary presentation of our general method, we may assume that $x \ge 3$ and $y \ge 3$.

From the above considerations we may assume that none of $x^2 + a$ and $y^2 + b$ is a perfect square. From (5.5.1) it follows that $x^2 + a = du^2$ and $y^2 + b = dv^2$ for some positive integers d, u, v. The last two equations can be written as

$$
x^2 - du^2 = -a \quad \text{and} \quad y^2 - dv^2 = -b \tag{5.5.3}
$$

which are general Pell's equations of the form $X^2 - dY^2 = N$, where $|N| \le 4$.

Define the set

$$
\mathcal{P}(a, b) = \{d \in \mathbb{Z} : d \text{ is nonsquare } \ge 2 \text{ and (5.5.3) are solvable}\} \tag{5.5.4}
$$

and for any d in $\mathcal{P}(a, b)$ consider the general solutions $(x(d), u(d))$ and $(y(d), v(d))$ to the equations (5.5.3) (see Chapter 4 for details).

We have $x^2(d) + a = du^2(d)$ and $y^2(d) + b = dv^2(d)$ hence

$$F(x(d), y(d), z) = \pm du(d)v(d). \tag{5.5.5}$$

Denote by $\mathcal{Z}_d$ the set of all integers z satisfying the equation (5.5.5).

The solutions to the equation (5.5.1) are $(x(d), y(d), z)$, where $d \in \mathcal{P}(a, b)$ and $z \in \mathcal{Z}_d$.

To illustrate this method let us consider the following concrete examples.

5.5.1   The Equation $x^2 + y^2 + z^2 + 2xyz = 1$

In the book [25] the above equation is solved in integers. Indeed, it is equivalent to

$$(x^2 - 1)(y^2 - 1) = (xy + z)^2, \tag{5.5.6}$$

an equation of the form (5.5.1), where $a = b = -1$ and $F(x, y, z) = xy + z$.

In this case $\mathcal{P}(-1, -1) = \{d > 0 : d \text{ nonsquare}\}$, as we have seen in Chapter 3.

Let $(s_l(d), t_l(d))_{l \geq 0}$ be the general solution to Pell's equation $s^2 - dt^2 = 1$. From the general method, it follows that the integral solutions to the given equation are

$$(\pm s_m(d), \pm s_n(d), -s_m(d)s_n(d) \pm dt_m(d)t_n(d)), \tag{5.5.7}$$

for all $m, n \geq 0$ and $d \in \mathcal{P}(-1, -1)$.

Using either of relations (3.2.2), (3.2.5), or (3.2.6), one can prove the following equalities

$$s_m(d)s_n(d) + dt_m(d)t_n(d) = s_{m+n}(d), \quad m, n \geq 0$$
$$s_m(d)s_n(d) - dt_m(d)t_n(d) = s_{m-n}(d), \quad m \geq n \geq 0.$$

The triples (5.5.7) become

$$\begin{aligned}(\pm s_m(d), \pm s_n(d), -s_{m+n}(d)), \quad m, n \geq 0 \\ (\pm s_m(d), \pm s_n(d), -s_{m-n}(d)), \quad m \geq n \geq 0,\end{aligned} \tag{5.5.8}$$

where the signs + and − correspond.

Given the symmetry of the equation in x, y, z, in order to obtain all of its solutions, we need to also consider the triples obtained from (5.5.8) by cyclic permutations. We mention that the solutions found in [216] are not complete.

Remark. The equation

$$x^2 + y^2 + z^2 + 2xyz = 1 \tag{5.5.9}$$

has an interesting history. Its geometric interpretation has been pointed out in [30], where it is shown that it reduces to finding all triangles whose angles have rational cosines.

The general solution in rational numbers of this equation is given [33]:

$$x = \frac{b^2 + c^2 - a^2}{2bc}, \quad y = \frac{a^2 + c^2 - b^2}{2ac}, \quad z = \frac{a^2 + b^2 - c^2}{2ab}.$$

In the paper [167] it is noted that apart from the trivial solutions $(\pm 1, 0, 0)$, $(0, \pm 1, 0)$, $(0, 0, \pm 1)$, all integral solutions to the equation (5.5.9) are given by the following rule: if p, q, r are any integers with greatest common divisor 1 such that one of them is equal to the sum of the other two and if $u \geq 1$ is any integer, then

$$x = \pm\mathrm{ch}(p\theta), \quad y = \pm\mathrm{ch}(q\theta), \quad z = \pm\mathrm{ch}(r\theta)$$

where $\theta = \ln\left(u + \sqrt{u^2 - 1}\right)$ and $u \geq 1$ is an arbitrary integer.

In the papers [148, 149] it is studied the more general Diophantine equation

$$x^2 + y^2 + z^2 + 2xyz = n. \tag{5.5.10}$$

It is proved that this equation has no solutions in integers if $n \equiv 3 \pmod 4$, $n \equiv 6 \pmod 8$, $n \equiv \pm 3 \pmod 9$, $n = 1 - 4k^2$ with $k \not\equiv 0 \pmod 4$ and k has no prime factors of the form $4j + 3$, or $n = 1 - 3k^2$ with $(k, 4) = 2$, $(k, 3) = 1$ and k has no prime factors of the form $3j + 2$. On the other hand, one solution to the equation (5.5.10) implies infinitely many such solutions, except possibly when n is a perfect square having no prime factors of the form $4j + 1$. Also, there are infinitely many solutions if $n = 2^r$ and r is odd, but only the solution $x = y = 0$, $z = 2^{\frac{r}{2}}$ and its cyclic permutations when r is even.

5.5.2   The Equation $x^2 + y^2 + z^2 - xyz = 4$

The problem of finding all triples of positive integers (x, y, z) with the property mentioned above appears in [7]. These triples were found by using our general method described at the beginning of this section. Indeed, writing the equation in the equivalent form

$$(x^2 - 4)(y^2 - 4) = (xy - 2z)^2$$

we note that in (5.5.1) we have $a = b = -4$ and $F(x, y, z) = xy - 2z$. Both of the equations (5.5.3) reduce to the special Pell's equation $s^2 - dt^2 = 4$, which was extensively discussed in Section 4.3.2. Let $(s_l(d), t_l(d))_{l \geq 0}$ be the general solution to the equation $s^2 - dt^2 = 4$ given in (4.4.2) or (4.4.5). From (5.5.5) we obtain

$$z_1 = \frac{1}{2}\big(s_m(d)s_n(d) + dt_m(d)t_n(d)\big) = s_{m+n}(d)$$

and

$$z_2 = \frac{1}{2}\big(s_m(d)s_n(d) - dt_m(d)t_n(d)\big) = s_{|m-n|}(d).$$

The general positive integral solutions to the equation $x^2 + y^2 + z^2 - xyz = 4$ are

$$(s_m(d), s_n(d), s_{m+n}(d)), \; m, n \geq 0 \text{ and } (s_m(d), s_n(d), s_{m-n}(d)), \; m \geq n \geq 0$$

along with the corresponding permutations.

5.5.3 The Equation $(x^2 + 1)(y^2 + 1) = z^2$

In order to solve this equation in positive integers x, y, z note that $a = b = 1$ and $F(x, y, z) = z$. The equations (5.5.3) become the negative Pell's equation $s^2 - dt^2 = -1$. As we have seen in Section 3.6 the set $\mathcal{P}(1, 1)$ is far from easy to describe. The general solution of this equation is

$$(s_m, s_n, dt_m t_n),$$

where $m, n \geq 0$ and $d \in \mathcal{P}(1, 1)$.

In a similar way one can solve the equations in $(k + 1)$ variables:

$$(x_1^2 \pm 1)(x_2^2 \pm 1) \ldots (x_k^2 \pm 1) = y^2 \tag{5.5.11}$$

for any choice of the signs $+$ and $-$.

5.5.4 The Equation $(x^2 - 1)(y^2 - 1) = (z^2 - 1)^2$

The problem of finding all solutions in positive integers to the equation

$$(x^2 - 1)(y^2 - 1) = (z^2 - 1)^2 \tag{5.5.12}$$

is still open [82]. Partial results concerning this equation were published in [120, 222, 227], and [228].

In what follows we will describe the set of solutions to the equation (5.5.12). Our description will show the complexity of the problem of finding all of its solutions. The equation (5.5.12) is of the type (5.5.1), where $a = b = -1$ and $F(x, y, z) = z^2 - 1$. By using the general method presented at the beginning of this section, we can describe the set of solutions to (5.5.12) in the following way: Fix a nonsquare $d \geq 2$ and consider the Pell's equation $s^2 - dt^2 = 1$. It is clear that the solutions to (5.5.12) are of the form $(s_m, s_n, z_{m,n})_{m,n \geq 0}$, where $(s_k, t_k)_{k \geq 0}$ is the general solution to the above Pell's equation, $z_{m,n} = \sqrt{1 + dt_m t_n}$, $m, n \geq 0$ and $1 + dt_m t_n$ is a perfect square.

Let

$$C_d = \{(s_m, s_n, z_{m,n}) : 1 + dt_m t_n \text{ is a square, } m, n \geq 0\}.$$

The set of all solutions to (5.5.12) is

$$C = \bigcup_{\substack{d \geq 2 \\ \sqrt{d} \notin \mathbb{Q}}} C_d.$$

Note that for all nonsquare $d \geq 2$, C_d contains the infinite family of solutions (s_m, s_m, s_m), $m \geq 0$, but this is far from determining all elements in C_d.

5.6 Other Equations with Infinitely Many Solutions

5.6.1 The Equation $x^2 + axy + y^2 = 1$

In the book [26] we determine all integers a for which the equation

$$x^2 + axy + y^2 = 1 \tag{5.6.1}$$

has infinitely many integer solutions (x, y). In case of solvability, we find all such solutions. Clearly, (5.6.1) is a special case of the general equation (4.8.1).

Theorem 5.6.1. *The equation (5.6.1) has infinitely many integer solutions if and only if $|a| \geq 2$.*

If $a = -2$, the solutions are $(m, m+1)$, $(m+1, m)$, $(-m, -m-1)$, $(-m-1, -m)$, $m \in \mathbb{Z}$.

If $a = 2$, the solutions are $(-m, m+1)$, $(m+1, -m)$, $(m, -m-1)$, $(-m-1, m)$, $m \in \mathbb{Z}$.

If $|a| > 2$, the solutions are

$$(-v_n, v_{n+1}), (v_n, -v_{n+1}), (-v_{n+1}, v_n), (v_{n+1}, -v_n), \tag{5.6.2}$$

where

$$v_n = \frac{1}{\sqrt{a^2 - 4}} \left[\left(\frac{a + \sqrt{a^2 - 4}}{2} \right)^n - \left(\frac{a - \sqrt{a^2 - 4}}{2} \right)^n \right], \quad n \geq 0. \tag{5.6.3}$$

Proof. Rewrite the given equation in the form

$$(2x + ay)^2 - (a^2 - 4)y^2 = 4. \tag{5.6.4}$$

If $|a| < 2$, then the curve described by (5.6.4) is an ellipse, and so only finitely many integer solutions occur.

If $|a| = 2$ then the equation (5.6.1) has infinitely many solutions, since it can be written as $(x \pm y)^2 = 1$.

If $|a| > 2$, then $a^2 - 4$ is not perfect square. In this case we have a special Pell's equation of the form

$$u^2 - (a^2 - 4)v^2 = 4. \tag{5.6.5}$$

This type of equations was extensively studied in Section 4.3.2.

Note that a nontrivial solution to (5.6.5) in $(a, 1)$. Using the formula (4.4.2) we obtain the general solution $(u_n, v_n)_{n \geq 1}$ to (5.6.5), where

$$u_n = \left(\frac{a + \sqrt{a^2 - 4}}{2} \right)^n + \left(\frac{a - \sqrt{a^2 - 4}}{2} \right)^n,$$

$$v_n = \frac{1}{\sqrt{a^2 - 4}} \left[\left(\frac{a + \sqrt{a^2 - 4}}{2} \right)^n - \left(\frac{a - \sqrt{a^2 - 4}}{2} \right)^n \right], \quad n \geq 1.$$

From formulas (4.4.3) the sequences $(u_n)_{n \geq 1}$, $(v_n)_{n \geq 1}$ satisfy the recursive system

$$\begin{cases} u_{n+1} = \dfrac{1}{2}[au_n + (a^2 - 4)v_n] \\[2mm] v_{n+1} = \dfrac{1}{2}(u_n + av_n), \quad n \geq 1. \end{cases} \tag{5.6.6}$$

From (5.6.4) it follows that the nontrivial integer solutions (x, y) to the equation (5.6.1) satisfy

$$2x + ay = \pm u_n \quad \text{and} \quad y = \pm v_n, \ n \geq 1 \tag{5.6.7}$$

where the signs $+$ and $-$ correspond.

If $2x + ay = u_n$ and $y = v_n$, then from (5.6.6) it follows that

$$x = \frac{1}{2}(u_n - av_n) = \frac{1}{4}[au_{n-1} + (a^2 - 4)v_{n-1} - au_{n-1} - a^2 v_{n-1}] = -v_{n-1}.$$

We obtain the solution $(-v_n, v_{n+1})_{n \geq 1}$. The choice $2x + ay = -u_n$ and $y = -v_n$ yield the solution $(v_n, -v_{n+1})_{n \geq 1}$ which in fact reflects the symmetry $(x, y) \to (-x, -y)$ of (5.6.1).

The last two solutions in (5.6.2) follow from the symmetry $(x, y) \to (y, x)$ of the equation (5.5.1).   □

Remark. In the case $a = -m$, where m is a positive integer, the equation (5.6.1) was solved in positive integers by using a complicated method involving planar transformations in [31, pp. 70–73].

5.6.2   The Equation $\dfrac{x^2 + 1}{y^2 + 1} = a^2 + 1$

We will prove that if a is any fixed positive integer, then there exist infinitely many pairs of positive integers (x, y) such that

$$\frac{x^2 + 1}{y^2 + 1} = a^2 + 1. \tag{5.6.8}$$

This means that the set $\{J_m = m^2 + 1 : \ m = 1, 2, \dots\}$ contains infinitely many pairs (J_x, J_y) such that $J_x = J_a J_y$.

The equation (5.6.8) is equivalent to the general Pell's equation

$$x^2 - (a^2 + 1)y^2 = a^2. \tag{5.6.9}$$

We notice that equation (5.6.9) has particular solutions $(a^2 - a + 1, a - 1)$ and $(a^2 + a + 1, a + 1)$. Let $(u_n, v_n)_{n \geq 0}$ be the general solution of its Pell's resolvent $u^2 - (a^2 + 1)v^2 = 1$. The fundamental solution to the Pell's resolvent equation is $(u_1, v_1) = (2a^2 + 1, 2a)$.

Following (4.1.3), we construct the sequences of solutions $(x_n, y_n)_{n \geq 0}$ and $(x'_n, y'_n)_{n \geq 0}$ to the equation (5.6.8):

$$\begin{cases} x_n = (a^2 - a + 1)u_n + (a - 1)(a^2 + 1)v_n \\ y_n = (a - 1)u_n + (a^2 - a + 1)v_n \end{cases} \tag{5.6.10}$$

and

$$\begin{cases} x'_n = (a^2 + a + 1)u_n + (a + 1)(a^2 + 1)v_n \\ y'_n = (a + 1)u_n + (a^2 + a + 1)v_n \end{cases} \tag{5.6.11}$$

We will show now that for all $a \geq 3$ the solutions $(x_n, y_n)_{n \geq 0}$, $(x'_n, y'_n)_{n \geq 0}$ are all distinct. In this respect, following the criterion given in Section 4.1 it suffices to see that at least one of the numbers

$$\frac{xx' - yy'd}{N} \quad \text{and} \quad \frac{yx' - xy'}{N}$$

is not an integer. Here (x, y) and (x', y') are solutions to the general Pell's equation $X^2 - dY^2 = N$. Indeed,

$$\frac{xx' - yy'd}{N} = \frac{(a^2 - a + 1)(a^2 + a + 1) - (a - 1)(a + 1)(a^2 + 1)}{a^2} =$$

$$= \frac{a^4 + a^2 + 1 - a^4 + 1}{a^2} = 1 + \frac{2}{a^2} \notin \mathbb{Z}.$$

Remarks. 1) The equation

$$(x^2 + 1)(y^2 + 1) = z^2 + 1 \tag{5.6.12}$$

was known even to Diophantus. It was him who pointed out the solutions $(k, 0, k)$, $(k, k \pm 1, k^2 \pm k + 1)$, where k is a positive integer and the signs + and − correspond.

The problem of finding all solutions to (5.6.12) in positive integers appears in [173]. Unfortunately, the solution presented there was incorrect.

2) It is clear that if $y = x + 1$, then $(x^2 + 1)(y^2 + 1) = (x^2 + 1)(x^2 + 2x + 2) = (x^2 + x + 1)^2 + 1$, hence the equation (5.6.12) has infinitely solutions (x, y, z), where x and y are consecutive positive integers.

In the case when x is fixed, the problem of finding infinitely many y and z satisfying (5.6.12) also appears in [205] and it is solved by using a suitable negative Pell's equation.

3) A weaker version of the same problem appears in [89] as follows: the sequence of numbers $J_l = l^2 + 1$, $l = 1, 2, \ldots$ contains an infinity of composite numbers $J_N = J_m \cdot J_n$. In fact, in the mentioned reference, for an arbitrary fixed m, only three pairs of corresponding n, N are found:

$$J_{m^2 - m + 1} = J_m \cdot J_{m-1}, \quad J_{m^2 + m + 1} = J_m \cdot J_{m+1} \quad \text{and} \quad J_{2m^2 + m} = J_m \cdot J_{2m^2}.$$

4) The equation (5.6.8) is completely solved in rational numbers in [50]. Its general solution is given by

$$x = f(a)$$

$$y = \pm a - 1 + \frac{2}{r^2(a) + 1} - \frac{2r(a)}{r^2(a) + 1} f(a),$$

where

$$f(a) = -a \pm \frac{(a^2 + 1)(r(a) \pm 1)^2}{r^2(a) + 2ar(a) - 1}$$

and $r(a)$ is any rational function of a.

5) A slightly modified equation is given by

$$\frac{x^2 - 1}{y^2 - 1} = z^2.$$ (5.6.13)

This equation can be solved completely [139]. Indeed, it is equivalent to

$$x^2 - (y^2 - 1)z^2 = 1.$$

It is not difficult to see that $(y, 1)$ is the fundamental solution to this equation and that all solutions are given by $(x_n, y, z_n)_{n \geq 0}$, where y is any integer greater than 1 and

$$\begin{cases} x_n = \dfrac{1}{2}\left[\left(y + \sqrt{y^2 - 1}\right)^n + \left(y - \sqrt{y^2 - 1}\right)^n\right] \\[4mm] z_n = \dfrac{1}{2\sqrt{y^2 - 1}}\left[\left(y + \sqrt{y^2 - 1}\right)^n - \left(y - \sqrt{y^2 - 1}\right)^n\right] \end{cases}$$

6) Another equation related to (5.6.12) is

$$\frac{x^2 + 1}{y^2 + 1} = z^2.$$ (5.6.14)

This equation can be solved completely as well. We write it under the form

$$x^2 - (y^2 + 1)z^2 = -1,$$

a negative Pell's equation with minimal solution $(y, 1)$. Using formulas (3.6.3) we obtain the general solution $(x_n, z_n)_{n \geq 0}$,

$$\begin{cases} x_n = \dfrac{1}{2}\left[\left(y + \sqrt{y^2 + 1}\right)^{2n+1} + \left(y - \sqrt{y^2 + 1}\right)^{2n+1}\right] \\[4mm] y_n = \dfrac{1}{2\sqrt{y^2 + 1}}\left[\left(y + \sqrt{y^2 + 1}\right)^{2n+1} - \left(y - \sqrt{y^2 + 1}\right)^{2n+1}\right] \end{cases}$$

All solutions to (5.6.14) are given by $(x_n, y, z_n)_{n \geq 0}$, where y is any positive integer.

5.6.3 The Equation $(x + y + z)^2 = xyz$

Generally, integer solutions to equations of three or more variables are given in various parametric forms. In this section we will construct different families of infinite nonzero integer solutions to the equation:

$$(x + y + z)^2 = xyz. \tag{5.6.15}$$

Following our paper [8] we will indicate a general method of generating such families of solutions. We start by performing the transformations

$$x = \frac{u + v}{2} + a, \quad y = \frac{u - v}{2} + a, \quad z = b \tag{5.6.16}$$

where a and b are nonzero integer parameters that will be determined in a convenient manner. The equation becomes

$$(u + 2a + b)^2 = \frac{b}{4}(u^2 - v^2) + abu + a^2b.$$

Imposing the conditions $2(2a + b) = ab$ and $b(b - 4) > 0$ yields the general Pell's equation

$$(b - 4)u^2 - bv^2 = 4[(2a + b)^2 - a^2b]. \tag{5.6.17}$$

The imposed conditions are equivalent to $(a - 2)(b - 4) = 8$, and $b < 0$ or $b > 4$. A simple case analysis shows that the only pairs of integers (a, b) satisfying them are: $(1, -4)$, $(3, 12)$, $(4, 8)$, $(6, 6)$, $(10, 5)$.

The following table contains the general Pell's equations (5.6.17) corresponding to the above pairs (a, b), their Pell's resolvents, both equations with their fundamental solutions.

(a, b)	General Pell's equation (5.6.17) and its fundamental solution	Pell's resolvent and its fundamental solution
$(1, -4)$	$2u^2 - v^2 = -8, \ (2, 4)$	$r^2 - 2s^2 = 1, \ (3, 2)$
$(3, 12)$	$2u^2 - 3v^2 = 216, \ (18, 12)$	$r^2 - 6s^2 = 1, \ (5, 2)$
$(4, 8)$	$u^2 - 2v^2 = 128, \ (16, 8)$	$r^2 - 2s^2 = 1, \ (3, 2)$
$(6, 6)$	$u^2 - 3v^2 = 216, \ (18, 6)$	$r^2 - 3s^2 = 1, \ (2, 1)$
$(10, 5)$	$u^2 - 5v^2 = 500, \ (25, 5)$	$r^2 - 5s^2 = 1, \ (9, 4)$

By using the formula (4.5.2) we obtain the following sequences of solutions to the equations (5.6.17):

$$u_m^{(1)} = 2r_m^{(1)} + 4s_m^{(1)}, \quad v_m^{(1)} = 4r_m^{(1)} + 4s_m^{(1)}$$

where $r_m^{(1)} + s_m^{(1)}\sqrt{2} = (3 + 2\sqrt{2})^m$, $m \geq 0$;

$$u_m^{(2)} = 18r_m^{(2)} + 36s_m^{(2)}, \quad v_m^{(2)} = 12r_m^{(2)} + 36s_m^{(2)}$$

where $r_m^{(2)} + s_m^{(2)}\sqrt{6} = (5 + 2\sqrt{6})^m$, $m \geq 0$;

$$u_m^{(3)} = 16r_m^{(3)} + 16s_m^{(3)}, \quad v_m^{(3)} = 8r_m^{(3)} + 16s_m^{(3)}$$

where $r_m^{(3)} + s_m^{(3)}\sqrt{2} = (3 + 2\sqrt{2})^m$, $m \geq 0$;

$$u_m^{(4)} = 18r_m^{(4)} + 18s_m^{(4)}, \quad v_m^{(4)} = 6r_m^{(4)} + 18s_m^{(4)}$$

where $r_m^{(4)} + s_m^{(4)}\sqrt{3} = (2 + \sqrt{3})^m$, $m \geq 0$;

$$u_m^{(5)} = 25r_m^{(5)} + 25s_m^{(5)}, \quad v_m^{(5)} = 5r_m^{(5)} + 25s_m^{(5)}$$

where $r_m^{(5)} + s_m^{(5)}\sqrt{5} = (9 + 4\sqrt{5})^m$, $m \geq 0$.

Formulas (5.6.16) yield the following five families of nonzero integer solutions to the equation (5.6.15):

$$x_m^{(1)} = 3r_m^{(1)} + 4s_m^{(1)} + 1, \quad y_m^{(1)} = -r_m^{(1)} + 1, \quad z_m^{(1)} = -4, \quad m \geq 0$$

$$x_m^{(2)} = 15r_m^{(2)} + 36s_m^{(2)} + 3, \quad y_m^{(2)} = 3r_m^{(2)} + 3, \quad z_m^{(2)} = 12, \quad m \geq 0$$

$$x_m^{(3)} = 12r_m^{(3)} + 16s_m^{(3)} + 4, \quad y_m^{(3)} = 4r_m^{(3)} + 4, \quad z_m^{(3)} = 8, \quad m \geq 0$$

$$x_m^{(4)} = 12r_m^{(4)} + 18s_m^{(4)} + 6, \quad y_m^{(4)} = 6r_m^{(4)} + 6, \quad z_m^{(4)} = 6, \quad m \geq 0$$

$$x_m^{(5)} = 15r_m^{(5)} + 25s_m^{(5)} + 10, \quad y_m^{(5)} = 10r_m^{(5)} + 10, \quad z_m^{(5)} = 5, \quad m \geq 0.$$

Remark. In the recent paper [78] the following approach to generates solutions to the equation (5.6.15) is indicated. Taking $z = kx - y$, for some integer k, our equation is equivalent to $y^2 - kxy + x(k+1)^2 = 0$, which is a quadratic equation in y, hence

$$y = \frac{1}{2}\left(kx \pm \sqrt{k^2x^2 - 4(k+1)^2x}\right).$$

Now, let $k^2x^2 - 4(k+1)^2x = a^2$, for some integer a. Treat this relation as a quadratic equation in x, we have

$$x = 2(k+1)^2 \pm \sqrt{4(k+1)^4 + k^2a^2}.$$

Again, consider $4(k+1)^4 + k^2a^2 = b^2$, for some integer b. Considering the last equation as $(2(k+1)^2)^2 + (ka)^2 = b^2$, which is a Pythagorean, we get the following two possible situations

$$\begin{cases} b = u^2 + v^2 \\ ka = u^2 - v^2 \\ (k+1)^2 = uv \end{cases} \quad \text{and} \quad \begin{cases} b = u^2 + v^2 \\ ka = 2uv \\ 2(k+1)^2 = u^2 - v^2 \end{cases}$$

where $u, v \in \mathbb{Z}$. Therefore, in order to generate solutions to equation (5.6.15), we start with two integers u, v such that uv or $\dfrac{u^2 - v^2}{2}$ is a perfect square $(k+1)^2$. Then, we find $a = \dfrac{u^2 - v^2}{k}$ or $a = \dfrac{2uv}{k}$, and $b = u^2 + v^2$. Finally, we obtain

$$x = 2(k+1)^2 \pm b, \quad y = \frac{1}{2}(kx \pm a), \quad z = kx - y.$$

Clearly, every pair (u, v) generates at most two values of k for each system considered above. Let us illustrate the method by the following special situation.

Example. Let $ka = \dfrac{3}{2}(k+1)^2$, $b = \dfrac{5}{2}(k+1)^2$, be the special solutions to the equation $(2(k+1)^2)^{\frac{3}{2}} + (ka)^2 = b^2$. Then we obtain from families (x, y) the solutions:

$$\left(\frac{9(k+1)^2}{2k^2}, \frac{6(k+1)^2}{k} \right), \quad \left(\frac{9(k+1)^2}{2k^2}, \frac{3(k+1)^2}{k} \right),$$

$$\left(-\frac{(k+1)^2}{2k^2}, \frac{(k+1)^2}{k} \right), \quad \left(-\frac{(k+1)^2}{2k^2}, -\frac{2(k+1)^2}{k} \right).$$

In order to get integer solutions, the only possibilities are $k = -3, -1, 1, 3$, giving the solutions $(0, 0, 0)$, $(18, 12, 6)$, $(8, 16, 8)$, $(-2, 2 - 4)$, $(-2, -8, 6)$.

5.6.4   The Equation $(x + y + z + t)^2 = xyzt$

Using the method described in Section 5.6.3 we will generate nine infinite families of positive integer solutions to the equation

$$(x + y + z + t)^2 = xyzt. \tag{5.6.18}$$

We will follow the paper [9].

The transformations

$$x = \frac{u + v}{2} + a, \quad y = \frac{u - v}{2} + a, \quad z = b, \quad t = c \tag{5.6.19}$$

where a, b, c are positive integers, bring the equation (5.6.18) to the form

$$(u + 2a + b + c)^2 = \frac{bc}{4}(u^2 - v^2) + abcu + a^2bc.$$

Setting the conditions $2(2a + b + c) = abc$ and $bc > 4$, we obtain the following general Pell's equation

$$(bc - 4)u^2 - bcv^2 = 4[(2a + b + c)^2 - a^2 bc]. \qquad (5.6.20)$$

There are nine triples (a, b, c) up to permutations satisfying the above conditions: (1,6,4), (1,10,3), (2,2,6), (3,4,2), (3,14,1), (5,2,3), (4,1,9), (7,1,6), (12,1,5).

The following table contains the general Pell's equations (5.6.20) corresponding to the above triples (a, b, c), their Pell's resolvent, both equations with their fundamental solutions.

(a, b, c)	General Pell's equation (5.6.20) and its fundamental solution	Pell's resolvent and its fundamental solution
$(1, 6, 4)$	$5u^2 - 6v^2 = 120,\ (12, 10)$	$r^2 - 30s^2 = 1,\ (11, 2)$
$(1, 10, 3)$	$13u^2 - 15v^2 = 390,\ (15, 13)$	$r^2 - 195s^2 = 1,\ (14, 1)$
$(2, 2, 6)$	$2u^2 - 3v^2 = 96,\ (12, 8)$	$r^2 - 6s^2 = 1,\ (5, 2)$
$(3, 4, 2)$	$u^2 - 2v^2 = 72,\ (12, 6)$	$r^2 - 2s^2 = 1,\ (3, 2)$
$(3, 14, 1)$	$5u^2 - 7v^2 = 630,\ (21, 15)$	$r^2 - 35s^2 = 1,\ (6, 1)$
$(4, 1, 9)$	$5u^2 - 9v^2 = 720,\ (42, 30)$	$r^2 - 45s^2 = 1,\ (161, 24)$
$(5, 2, 3)$	$u^2 - 3v^2 = 150,\ (15, 5)$	$r^2 - 3s^2 = 1,\ (2, 1)$
$(7, 1, 6)$	$u^2 - 3v^2 = 294,\ (21, 7)$	$r^2 - 3s^2 = 1,\ (2, 1)$
$(12, 1, 5)$	$u^2 - 5v^2 = 720,\ (30, 6)$	$r^2 - 5s^2 = 1,\ (9, 4)$

By using the formula (4.4.2) we obtain the following sequences of solutions to equations (5.6.20):

$$u_m^{(1)} = 12r_m^{(1)} + 60s_m^{(1)}, \quad v_m^{(1)} = 10r_m^{(1)} + 60s_m^{(1)},$$

where $r_m^{(1)} + s_m^{(1)}\sqrt{30} = (11 + 2\sqrt{30})^m$, $m \geq 0$;

$$u_m^{(2)} = 15r_m^{(2)} + 195s_m^{(2)}, \quad v_m^{(2)} = 13r_m^{(2)} + 195s_m^{(2)},$$

where $r_m^{(2)} + s_m^{(2)}\sqrt{195} = (14 + \sqrt{195})^m$, $m \geq 0$;

$$u_m^{(3)} = 12r_m^{(3)} + 24s_m^{(3)}, \quad v_m^{(3)} = 8r_m^{(3)} + 24s_m^{(3)},$$

where $r_m^{(3)} + s_m^{(3)}\sqrt{6} = (5 + 2\sqrt{6})^m$, $m \geq 0$;

$$u_m^{(4)} = 12r_m^{(4)} + 12s_m^{(4)}, \quad v_m^{(4)} = 6r_m^{(4)} + 12s_m^{(4)},$$

where $r_m^{(4)} + s_m^{(4)}\sqrt{2} = (3 + 2\sqrt{2})^m$, $m \geq 0$;

$$u_m^{(5)} = 21r_m^{(5)} + 105s_m^{(5)}, \quad v_m^{(5)} = 15r_m^{(5)} + 105s_m^{(5)},$$

where $r_m^{(5)} + s_m^{(5)}\sqrt{35} = (6 + \sqrt{35})^m$, $m \geq 0$;

$$u_m^{(6)} = 42r_m^{(6)} + 270s_m^{(6)}, \quad v_m^{(6)} = 30r_m^{(6)} + 210s_m^{(6)},$$

where $r_m^{(6)} + s_m^{(6)}\sqrt{45} = (161 + 24\sqrt{45})^m$, $m \geq 0$;

$$u_m^{(7)} = 15r_m^{(7)} + 15s_m^{(7)}, \quad v_m^{(7)} = 5r_m^{(7)} + 15s_m^{(7)},$$

where $r_m^{(7)} + s_m^{(7)}\sqrt{3} = (2 + \sqrt{3})^m$, $m \geq 0$;

$$u_m^{(8)} = 21r_m^{(8)} + 21s_m^{(8)}, \quad v_m^{(8)} = 7r_m^{(8)} + 21s_m^{(8)},$$

where $r_m^{(8)} + s_m^{(8)}\sqrt{3} = (2 + \sqrt{3})^m$, $m \geq 0$;

$$u_m^{(9)} = 30r_m^{(9)} + 30s_m^{(9)}, \quad v_m^{(9)} = 6r_m^{(9)} + 30s_m^{(9)},$$

where $r_m^{(9)} + s_m^{(9)}\sqrt{5} = (9 + 4\sqrt{5})^m$, $m \geq 0$.

Formulas (5.6.19) yield the following nine families of positive integers solutions to the equation (5.6.18):

$$x_m^{(1)} = 11r_m^{(1)} + 60s_m^{(1)} + 1, \quad y_m^{(1)} = r_m^{(1)} + 1, \quad z_m^{(1)} = 6, \quad t_m^{(1)} = 4$$

$$x_m^{(2)} = 14r_m^{(2)} + 195s_m^{(2)} + 1, \quad y_m^{(2)} = r_m^{(2)} + 1, \quad z_m^{(2)} = 10, \quad t_m^{(2)} = 3$$

$$x_m^{(3)} = 10r_m^{(3)} + 24s_m^{(3)} + 2, \quad y_m^{(3)} = 2r_m^{(3)} + 2, \quad z_m^{(3)} = 2, \quad t_m^{(3)} = 6$$

$$x_m^{(4)} = 12r_m^{(4)} + 12s_m^{(4)} + 3, \quad y_m^{(4)} = 3r_m^{(4)} + 3, \quad z_m^{(4)} = 4, \quad t_m^{(4)} = 2$$

$$x_m^{(5)} = 18r_m^{(5)} + 105s_m^{(5)} + 3, \quad y_m^{(5)} = 3r_m^{(5)} + 3, \quad z_m^{(5)} = 14, \quad t_m^{(5)} = 1$$

$$x_m^{(6)} = 36r_m^{(6)} + 240s_m^{(6)} + 4, \quad y_m^{(6)} = 6r_m^{(6)} + 30s_m^{(6)} + 4, \quad z_m^{(6)} = 1, \quad t_m^{(6)} = 9$$

$$x_m^{(7)} = 10r_m^{(7)} + 15s_m^{(7)} + 5, \quad y_m^{(7)} = 5r_m^{(7)} + 5, \quad z_m^{(7)} = 2, \quad t_m^{(7)} = 3$$

$$x_m^{(8)} = 14r_m^{(8)} + 21s_m^{(8)} + 7, \quad y_m^{(8)} = 7r_m^{(8)} + 7, \quad z_m^{(8)} = 1, \quad t_m^{(8)} = 6$$

$$x_m^{(9)} = 18r_m^{(9)} + 30s_m^{(9)} + 12, \quad y_m^{(9)} = 12r_m^{(9)} + 12, \quad z_m^{(9)} = 1, \quad t_m^{(9)} = 5.$$

Remarks. 1) In [194] only solution $(x_m^{(7)}, y_m^{(7)}, z_m^{(7)}, t_m^{(7)})$ is found.

2) Note the atypical form of solution $(x_m^{(6)}, y_m^{(6)}, z_m^{(6)}, t_m^{(6)})_{m \geq 0}$.

5.6.5   The Equation $(x + y + z + t)^2 = xyzt + 1$

The equation

$$(x + y + z + t)^2 = xyzt + 1 \qquad (5.6.21)$$

is considered in the paper [79], where the method to generate families of solutions is similar to the one described in the previous section. Introduction of the linear transformations

$$x = u + v + a, \ y = u - v + a, \ z = b, \ t = c, \qquad (5.6.22)$$

where a, b, c are positive integers, leads (5.6.21) to the form

$$(bc - 4)u^2 - bcv^2 = (2a + b + c)^2 - a^2bc - 1, \qquad (5.6.23)$$

in which $bc > 4$ and $2(2a+b+c) = abc$. There are six triples (a, b, c) satisfying the above conditions, namely $(5,2,3)$, $(7,1,6)$, $(12,1,5)$, $(1,10,3)$, $(3,14,1)$, $(4,1,9)$. The following table contains the general Pell's equations (5.6.23) corresponding to the above triples (a, b, c), their Pell's resolvent, both equations with their fundamental solutions.

(a, b, c)	General Pell's equation (5.6.23) and its fundamental solution	Pell's resolvent and its fundamental solution
$(5, 2, 3)$	$u^2 - 3v^2 = 37, \ (7, 2)$	$r^2 - 3s^2 = 1, \ (2, 1)$
$(7, 1, 6)$	$u^2 - 3v^2 = 73, \ (10, 3)$	$r^2 - 3s^2 = 1, \ (2, 1)$
$(12, 1, 5)$	$u^2 - 5v^2 = 179, \ (28, 11)$	$r^2 - 5s^2 = 1, \ (9, 4)$
$(1, 10, 3)$	$13u^2 - 15v^2 = 97, \ (7, 6)$	$r^2 - 195s^2 = 1, \ (14, 1)$
$(3, 14, 1)$	$5u^2 - 7v^2 = 157, \ (10, 7)$	$r^2 - 35s^2 = 1, \ (6, 1)$
$(4, 1, 9)$	$5u^2 - 9v^2 = 179, \ (50, 37)$	$r^2 - 45s^2 = 1, \ (161, 24)$

In view of the formula (4.4.2), the following sequences are six families of positive integer solutions to the corresponding general Pell's equations (5.6.22):

$$u_m^{(1)} = 7r_m^{(1)} + 6s_m^{(1)}, \ v_m^{(1)} = 2r_m^{(1)} + 7s_m^{(1)},$$

where $r_m^{(1)} + s_m^{(1)}\sqrt{3} = (2 + \sqrt{3})^m$, $m \geq 0$.

$$u_m^{(2)} = 10r_m^{(2)} + 9s_m^{(2)}, \ v_m^{(2)} = 3r_m^{(2)} + 10s_m^{(2)},$$

where $r_m^{(2)} + s_m^{(2)}\sqrt{3} = (2 + \sqrt{3})^m$, $m \geq 0$.

$$u_m^{(3)} = 28r_m^{(3)} + 55s_m^{(3)}, \quad v_m^{(3)} = 11r_m^{(3)} + 28s_m^{(3)},$$

where $r_m^{(3)} + s_m^{(3)}\sqrt{5} = (9 + 4\sqrt{5})^m$, $m \geq 0$.

$$u_m^{(4)} = 7r_m^{(4)} + 90s_m^{(4)}, \quad v_m^{(4)} = 6r_m^{(4)} + 91s_m^{(4)},$$

where $r_m^{(4)} + s_m^{(4)}\sqrt{195} = (14 + \sqrt{195})^m$, $m \geq 0$.

$$u_m^{(5)} = 10r_m^{(5)} + 49s_m^{(5)}, \quad v_m^{(5)} = 7r_m^{(5)} + 50s_m^{(5)},$$

where $r_m^{(5)} + s_m^{(5)}\sqrt{35} = (6 + \sqrt{35})^m$, $m \geq 0$.

$$u_m^{(6)} = 50r_m^{(6)} + 333s_m^{(6)}, \quad v_m^{(6)} = 37r_m^{(6)} + 250s_m^{(6)},$$

where $r_m^{(6)} + 3s_m^{(6)}\sqrt{5} = (161 + 72\sqrt{5})^m$, $m \geq 0$.

Formulas (5.6.22) yield the following six families of positive integers solutions to the equation (5.6.21):

$$x_m^{(1)} = 9r_m^{(1)} + 13s_m^{(1)} + 5, \ y_m^{(1)} = 5r_m^{(1)} - s_m^{(1)} + 5, \ z_m^{(1)} = 2, \ t_m^{(1)} = 3$$

$$x_m^{(2)} = 13r_m^{(2)} + 19s_m^{(2)} + 7, \ y_m^{(2)} = 7r_m^{(2)} - s_m^{(2)} + 7, \ z_m^{(2)} = 1, \ t_m^{(2)} = 6$$

$$x_m^{(3)} = 39r_m^{(3)} + 83s_m^{(3)} + 12, \ y_m^{(3)} = 17r_m^{(3)} + 27s_m^{(3)} + 12, \ z_m^{(3)} = 1, \ t_m^{(3)} = 5$$

$$x_m^{(4)} = 13r_m^{(4)} + 181s_m^{(4)} + 1, \ y_m^{(4)} = r_m^{(4)} - s_m^{(4)} + 1, \ z_m^{(4)} = 10, \ t_m^{(4)} = 3$$

$$x_m^{(5)} = 17r_m^{(5)} + 99s + 3, \ y_m^{(5)} = 3r_m^{(5)} - s_m^{(5)} + 3, \ z_m^{(5)} = 14, \ t_m^{(5)} = 1$$

$$x_m^{(6)} = 87r_m^{(6)} + 583s_m^{(6)} + 4, \ y_m^{(6)} = 13r_m^{(6)} + 83s_m^{(6)} + 4, \ z_m^{(6)} = 1, \ t_m^{(6)} = 9.$$

5.6.6 The Equation $x^3 + y^3 + z^3 + t^3 = n$

We will prove that if the equation

$$x^3 + y^3 + z^3 + t^3 = n \tag{5.6.24}$$

has an integral solution (a, b, c, d) such that $a \neq b$ or $c \neq d$ and $-(a+b)(c+d) > 0$ is not a perfect square, then it has infinitely many integral solutions.

For this, let us perform the transformations:

$$x = X + a, \quad y = -X + b, \quad z = Y + c, \quad t = -Y + d.$$

Then $(a + b)X^2 + (a^2 - b^2)X + (c + d)Y^2 + (c^2 - d^2)Y = 0$. The last equation is equivalent to

$$(a+b)\left(X+\frac{a-b}{2}\right)^2 + (c+d)\left(Y+\frac{c-d}{2}\right)^2$$

$$= \frac{(a+b)(a-b)^2}{4} + \frac{(c+d)(c-d)^2}{4}. \qquad (5.6.25)$$

From the hypothesis, (5.6.25) is a general Pell's equation:

$$AU^2 - BV^2 = C \qquad (5.6.26)$$

where $A = a+b$, $B = -(c+d)$, $C = \frac{1}{4}[(a+b)(a-b)^2 + (c+d)(c-d)^2]$ and $U = X + \frac{a-b}{2}$, $V = Y + \frac{c-d}{2}$.

We note that $(U_0, V_0) = \left(\frac{a-b}{2}, \frac{c-d}{2}\right)$ satisfies the equation (5.6.26) and consider the Pell's resolvent $r^2 - Ds^2 = 1$, where $D = -(a+b)(c+d)$, with the general solution $(r_m, s_m)_{m\geq 0}$. From the formula (4.5.2), we obtain the solutions $(U_m, V_m)_{m\geq 0}$ where

$$U_m = \frac{a-b}{2}r_m - (c+d)\frac{c-d}{2}s_m$$

$$V_m = \frac{c-d}{2}r_m + (a+b)\frac{a-b}{2}s_m.$$

It follows that

$$X_m = \frac{a-b}{2}r_m - \frac{c^2-d^2}{2}s_m - \frac{a-b}{2}$$

$$Y_m = \frac{c-d}{2}r_m + \frac{a^2-b^2}{2}s_m - \frac{c-d}{2}.$$

From these formulas we generate an infinite family of solutions $(x_m, y_m, z_m, t_m)_{m\geq 0}$ to the equation (5.6.24):

$$\begin{cases} x_m = \dfrac{a-b}{2}r_m - \dfrac{c^2-d^2}{2}s_m + \dfrac{a+b}{2} \\[2mm] y_m = -\dfrac{a-b}{2}r_m + \dfrac{c^2-d^2}{2}s_m + \dfrac{a+b}{2} \\[2mm] z_m = \dfrac{c-d}{2}r_m + \dfrac{a^2-b^2}{2}s_m + \dfrac{c+d}{2} \\[2mm] t_m = -\dfrac{c-d}{2}r_m - \dfrac{a^2-b^2}{2}s_m + \dfrac{c+d}{2}. \end{cases} \qquad (5.6.27)$$

Remarks. 1) The main idea of the approach described above comes from [150] and all computations are given in [10].

2) A special case of equation (5.6.24) appears in the book [24]: Prove that the equation

$$x^3 + y^3 + z^3 + t^3 = 1999$$

has infinitely many integral solutions (1999 Bulgarian Mathematical Olympiad). In this case, one simple solution to the given equation is $(a, b, c, d) = (10, 10, -1, 0)$. By using formulas (5.6.27), we obtain the following infinite family of solutions:

$$(x_m, y_m, z_m, t_m) = \left(-\frac{1}{2}s_m + 10, \frac{1}{2}s_m + 10, -\frac{1}{2}(r_m + 1), \frac{1}{2}(r_m - 1) \right),$$

where $r_m + s_m \sqrt{20} = (9 + 2\sqrt{20})^m$, $m \geq 0$. It is not difficult to see that the integers r_m are all odd and that s_m are all even.

Chapter 6
Diophantine Representations of Some Sequences

In 1900, David Hilbert asked for an algorithm to decide whether a given Diophantine equation is solvable or not and put this problem tenth in his famous list of 23.

In 1970, it was proved that such an algorithm cannot exist, i.e., the problem is recursively undecidable. Proof was supplied by Yu. V. Matiyasevich [133], heavily leaning on results arrived at by M. Davis, J. Robinson, and H. Putnam [60]. This was accomplished by proving that any enumerable set $A \subseteq \mathbb{N} = \{0, 1, 2, \dots\}$ can be represented in the following form: There exists a polynomial $p(x, x_1, \dots, x_n)$ with $n \geq 0$ such that $a \in A$ if and only if $p(a, x_1, \dots, x_n) = 0$ is solvable for particular nonnegative integers $x_1, \dots, x_n$, i.e.,

$$a \in A \Leftrightarrow \exists\, x_1, \dots, x_n \geq 0 : p(a, x_1, \dots, x_n) = 0.$$

Therefore, the set A equals the set of parameters for which the equation $p = 0$ is solvable. Employing an idea of H. Putnam [178] this can be reformulated as follows. If $q(x, x_1, \dots, x_n) = x(1 - p(x, x_1, \dots, x_n)^2)$, then A equals the set of positive values of q, where its variables range over the nonnegative integers. Among the recursively enumerable sets there are many for which such representation is surprising. We will name some examples which are of importance in number theory.

(1) The primes and their recursively enumerable subsets, most outstanding Fermat-, Mersenne-, and Twin-primes.
(2) The set of partial denominators of the continued fraction expansion of numbers as e, π and $\sqrt[3]{2}$. (Whereas for e this is known to equal $\{1\} \cup \{2, 4, 6, \dots\}$, there is only computer-based research regarding the other numbers.)

In this chapter we will introduce a Diophantine representation concept for sequences of integers that refines the idea of Diophantine set. This concept proves helpful in solving several types of Diophantine equations.

© Springer Science+Business Media New York 2015
T. Andreescu, D. Andrica, *Quadratic Diophantine Equations*,
Developments in Mathematics 40, DOI 10.1007/978-0-387-54109-9_6

6.1   Diophantine r-Representable Sequences

The sequence $(x_m)_{m \geq 1}$ is *Diophantine r-representable* if there exists a sequence $(P_n)_{n \geq 1}$ of polynomials of degree r, $P_n \in \mathbb{Z}[X_1, \ldots, X_r]$, such that for any positive integer n the following equality holds:

$$P_n(x_{n-r+1}, \ldots, x_n) = 0. \tag{6.1.1}$$

This means that the sequence $(x_m)_{m \geq 1}$ has the above property if and only if among the solutions to the Diophantine equation

$$P_n(y_1, y_2, \ldots, y_r) = 0$$

there are some for which $y_1^{(n)} = x_{n-r+1}$, $y_2^{(n)} = x_{n-r+2}, \ldots, y_r^{(n)} = x_n$, for all positive integers n.

The main result of this section is that any sequence defined by a linear recurrence of order r is Diophantine r-representable. Our approach follows the method given in [27] and [47] (see also [28] in the case $r = 2$).

Consider the sequence $(x_n)_{n \geq 1}$ defined recursively by

$$\begin{cases} x_i = \alpha_i, \ i = 1, 2, \ldots, r \\ x_n = \sum_{k=1}^{r} a_k x_{n-r-1+k}, \ n \geq r+1 \end{cases} \tag{6.1.2}$$

where $\alpha_1, \alpha_2, \ldots, \alpha_r$ and $a_1, a_2, \ldots, a_r$ are integers with $a_1 \neq 0$.

For $n \geq r$, let

$$D_n = \det \begin{bmatrix} x_{n-r+1} & x_{n-r+2} & \cdots & x_{n-1} & x_n \\ x_{n-r+2} & x_{n-r+3} & \cdots & x_n & x_{n+1} \\ \cdots & \cdots & \cdots & \cdots & \cdots \\ x_n & x_{n+1} & \cdots & x_{n+r-2} & x_{n+r-1} \end{bmatrix} \tag{6.1.3}$$

Lemma 6.1.1. *For all integers $n \geq r$, the following equality holds:*

$$D_n = (-1)^{(r-1)(n-r)} a_1^{n-r} D_r. \tag{6.1.4}$$

Proof. Following the method of [104, 135] and [202] we introduce the matrix

$$A_n = \begin{bmatrix} x_{n-r+1} & x_{n-r+2} & \cdots & x_{n-1} & x_n \\ x_{n-r+2} & x_{n-r+3} & \cdots & x_n & x_{n+1} \\ \cdots & \cdots & \cdots & \cdots & \cdots \\ x_{n-1} & x_n & \cdots & x_{n+r-3} & x_{n+r-2} \\ x_n & x_{n+1} & \cdots & x_{n+r-2} & x_{n+r-1} \end{bmatrix}.$$

It is easy to see that

$$
A_{n+1} =
\begin{bmatrix}
0 & 1 & 0 & 0 & \cdots & 0 & 0 & 0 \\
0 & 0 & 1 & 0 & \cdots & 0 & 0 & 0 \\
\hdotsfor{8} \\
0 & 0 & 0 & 0 & \cdots & 0 & 1 & 0 \\
0 & 0 & 0 & 0 & \cdots & 0 & 0 & 1 \\
a_1 & a_2 & a_3 & a_4 & \cdots & a_{r-2} & a_{r-1} & a_r
\end{bmatrix}
\cdot A_n
$$

and so that

$$
A_n =
\begin{bmatrix}
0 & 1 & 0 & 0 & \cdots & 0 & 0 & 0 \\
0 & 0 & 1 & 0 & \cdots & 0 & 0 & 0 \\
\hdotsfor{8} \\
0 & 0 & 0 & 0 & \cdots & 0 & 1 & 0 \\
0 & 0 & 0 & 0 & \cdots & 0 & 0 & 1 \\
a_1 & a_2 & a_3 & a_4 & \cdots & a_{r-2} & a_{r-1} & a_r
\end{bmatrix}^{n-r}
\cdot A_r.
\tag{6.1.5}
$$

Passing to determinants in (6.1.5), we obtain $((-1)^{r-1}a_1)^{n-r}D_r = D_n$ for $n \geq r$, that is the relation (6.1.4). $\qquad\square$

Theorem 6.1.2. *Any sequence defined by a linear recurrence of order r is Diophantine r-representable.*

Proof. Consider the sequence $(x_n)_{n \geq 1}$ defined by (6.1.2) and let $P_n \in \mathbb{Z}[X_1, \ldots, X_r]$ be the polynomial given by

$$
P_n(y_1, \ldots, y_r) = F_r(y_1, \ldots, y_r) - (-1)^{(r-1)(n-r)}a_1^{n-r}F_r(\alpha_1, \ldots, \alpha_r)
\tag{6.1.6}
$$

where $F_r \in \mathbb{Z}[X_1, \ldots, X_r]$ is obtained from the determinant (6.1.3) and the recursive relation (6.1.2).

From the relation (6.1.4) it follows that for all $n \geq r$ the following equalities hold

$$
P_n(x_{n-r+1}, \ldots, x_n) = F_r(x_{n-r+1}, \ldots, x_n) - (-1)^{(r-1)(n-r)}a_1^{n-r}F_r(\alpha_1, \ldots, \alpha_r)
$$

$$
= D_n - (-1)^{(r-1)(n-r)}a_1^{n-r}D_r = 0,
$$

i.e., the sequence $(x_n)_{n \geq 1}$ is Diophantine r-representable. $\qquad\square$

Remarks. 1) When $r = 2$, the polynomial F_r in (6.1.6) is given by

$$
F_2(x, y) = x^2 - a_2 xy - a_1 y^2
\tag{6.1.7}
$$

and it follows that, for the sequence $(x_n)_{n \geq 1}$ defined by

$$
\begin{cases}
x_1 = \alpha_1, \ x_2 = \alpha_2 \\
x_n = a_1 x_{n-2} + a_2 x_{n-1}, \ n \geq 3
\end{cases}
\tag{6.1.8}
$$

the relation $F_2(x_{n-1}, x_n) = (-1)^n a_1^{n-2} F_2(\alpha_1, \alpha_2)$ holds, i.e.,

$$x_n^2 - a_2 x_{n-1} x_n - a_1 x_{n-1}^2 = (-1)^n a_1^{n-2} (\alpha_2^2 - a_2 \alpha_1 \alpha_2 - a_1 \alpha_1^2). \qquad (6.1.9)$$

The relation (6.1.9) is the first relation of [35] and [36].

2) In the particular case $r = 3$, after elementary calculation, we obtain

$$\begin{aligned} F_3(x, y, z) = & -x^3 (a_1 + a_2 a_3) y^3 - a_1^2 z^3 + 2 a_3 x^2 y + a_2 x^2 z \\ & -(a_2^2 + a_1 a_3) y^2 z - (a_3^2 - a_2) x y^2 \\ & -a_1 a_3 x z^2 - 2 a_1 a_2 y z^2 + (3 a_1 - a_2 a_3) x y z, \end{aligned}$$

hence we get that, for the linear recurrence

$$\begin{cases} x_1 = \alpha_1, \ x_2 = \alpha_2, \ x_3 = \alpha_3 \\ x_n = a_1 x_{n-2} + a_2 x_{n-2} + a_3 x_{n-1}, \ n \geq 4 \end{cases} \qquad (6.1.10)$$

the relation

$$F_3(x_{n-2}, x_{n-1}, x_n) = a_1^{n-3} F_3(\alpha_1, \alpha_2, \alpha_3) \qquad (6.1.11)$$

is true.

3) If in the proof of Theorem 6.1.2, the equation $P_n(x_{n-r+1}, \ldots, x_n) = 0$ can be solved with respect to x_n, then x_n can be written as a function in $r - 1$ variables $x_{n-r+1}, \ldots, x_{n-1}$.

4) Note that the polynomial $F_r \in \mathbb{Z}[X_1, \ldots, X_r]$ can be viewed as an "invariant" to the sequence $(x_n)_{n \geq 1}$ defined by (6.1.2).

5) For additional informations about the special case $r = 2$ we refer to [92].

6.2 A Property of Some Special Sequences

If $a_1 = a_2 = 1$ and $\alpha_1 = \alpha_2 = 1$, then (6.1.8) defines the Fibonacci sequence $(F_n)_{n \geq 1}$ (see [105, 114, 187]) and [217] for many interesting properties). From (6.1.9) we obtain

$$F_n^2 - F_n F_{n-1} - F_{n-1}^2 = (-1)^{n-1}. \qquad (6.2.1)$$

If $a_1 = a_2 = 1$ and $\alpha_1 = 1$, $\alpha_2 = 3$, then (6.1.8) defines the Lucas sequence $(L_n)_{n \geq 1}$ (see [105]) and from (6.1.9) it follows that

$$L_n^2 - L_n L_{n-1} - L_{n-1}^2 = 5(-1)^n. \qquad (6.2.2)$$

If $a_1 = 1$, $a_2 = 2$ and $\alpha_1 = 1$, $\alpha_2 = 3$, then (6.1.8) gives the Pell sequence $(P_n)_{n \geq 1}$ (see [106]) and the relation (6.1.9) becomes

$$P_n^2 - 2P_n P_{n-1} - P_{n-1}^2 = (-1)^{n-1}. \tag{6.2.3}$$

From the relations (6.2.1), (6.2.2), (6.2.3) we deduce

$$F_n = \frac{1}{2}\left(F_{n-1} + \sqrt{5F_{n-1}^2 + 4(-1)^{n-1}}\right) \tag{6.2.4}$$

$$L_n = \frac{1}{2}\left(L_{n-1} + \sqrt{5L_{n-1}^2 + 20(-1)^n}\right) \tag{6.2.5}$$

$$P_n = P_{n-1} + \sqrt{2P_{n-1}^2 + (-1)^{n-1}}. \tag{6.2.6}$$

These identities give the possibility for writing computer programs that facilitate the computation of the terms of each of the three sequences $(F_n)_{n \geq 1}$, $(L_n)_{n \geq 1}$, $(P_n)_{n \geq 1}$.

In [183] it is given a method for obtaining the relation (6.2.4) by using hyperbolic functions. Similar results are also presented in [90].

Proposition 6.2.1. *If the sequence $(x_n)_{n \geq 1}$ is given by (6.1.8), then for all integers $n \geq 3$, the integer*

$$(a_2^2 + 4a_1)x_{n-1}^2 + 4(-1)^{n-1}a_1^{n-2}(a_1\alpha_1^2 + a_2\alpha_1\alpha_2 - \alpha_2^2)$$

is a perfect square.

Proof. From (6.1.9) we obtain

$$(a_2^2 + 4a_1)x_{n-1}^2 + 4(-1)^{n-1}a_1^{n-2}(a_1\alpha_1^2 + a_2\alpha_1\alpha_2 - \alpha_2^2) = (2x_n - a_2 x_{n-1})^2$$

which finishes the proof. $\qquad\square$

Proposition 6.2.2. *Let α_1, α_2 and k be nonzero integers. The general Pell's equations*

$$x^2 - (k^2 + 4)y^2 = 4(\alpha_1^2 + k\alpha_1\alpha_2 - \alpha_2^2)$$

and

$$(k^2 + 4)u^2 - v^2 = 4(\alpha_1^2 + k\alpha_1\alpha_2 - \alpha_2^2)$$

are solvable.

Proof. In (6.1.8) consider $a_1 = 1$ and $a_2 = k$. From Proposition 6.2.1 it follows that $(x, y) = (2x_n - kx_{n-1}, x_{n-1})$ is a solution to the first equation whenever n is odd. If n is even, then $(u, v) = (x_{n-1}, 2x_n - kx_{n-1})$ is a solution to the second equation. $\quad\square$

Remark. Note that the first equation in Proposition 6.2.2 has solution $(2\alpha_1 + k\alpha_2, \alpha_2)$. From Theorem 4.5.1 it follows that it has infinitely many integral solutions.

Similarly, the second equation in Proposition 6.2.2 has solution $(k\alpha_1 + (k^2 + 2)\alpha_2, \alpha_1 + k\alpha_2)$, and by applying Theorem 4.5.1 we deduce that it has infinitely many integral solutions.

6.3   The Equations $x^2 + axy + y^2 = \pm 1$

The result in Theorem 5.6.1 shows that if $|a| > 2$ the pairs $(-v_n, v_{n+1})$, $(v_n, -v_{n+1})$, $(-v_{n+1}, v_n)$, $(v_{n+1}, -v_n)$ of consecutive terms in the sequence given by

$$ v_n = \frac{1}{\sqrt{a^2 - 4}} \left[\left(\frac{a + \sqrt{a^2 - 4}}{2} \right)^n - \left(\frac{a - \sqrt{a^2 - 4}}{2} \right)^n \right] $$

can be characterized as solutions to the equation $x^2 + ax + y^2 = 1$.

On the other hand, the sequence $(v_n)_{n \geq 0}$ satisfies the linear recurrence of order 2

$$ v_{n+1} = av_n - v_{n-1}, \ n \geq 1, \text{ where } v_0 = 0 \text{ and } v_1 = 1. $$

Therefore the solutions to the discussed equation consists of all pairs of consecutive terms in a sequence defined by a second order recursive linear relation. In what follows, we will study the equation

$$ x^2 + axy + y^2 = -1, \tag{6.3.1} $$

which is also a special case of (4.8.1).

Theorem 6.3.1. *The equation (6.3.1) is solvable in integers if and only if $a = \pm 3$. If $a = -3$, then the solutions are*

$$ (-F_{2n-1}, -F_{2n+1}), \ (-F_{2n+1}, -F_{2n-1}), $$
$$ (F_{2n-1}, F_{2n+1}), \ (F_{2n+1}, F_{2n-1}), \ n \geq 1. $$

If $a = 3$, then the solutions are

$$ (-F_{2n-1}, F_{2n+1}), \ (-F_{2n+1}, F_{2n-1}), $$
$$ (F_{2n-1}, -F_{2n+1}), \ (F_{2n+1}, -F_{2n-1}), \ n \geq 1, $$

where $(F_m)_{m \geq 1}$ is the Fibonacci sequence.

Proof. First consider $a < 0$. If there is a solution (x, y), then $xy > 0$. Therefore, we may assume that $x > 0$, $y > 0$ and we may consider that x is minimal.

If $a \neq -3$, then $x \neq y$, for otherwise $(a + 2)x^2 = -1$, which is impossible, because $a + 2 \neq -1$. We have

$$0 = x^2 + axy + y^2 + 1 = (x + ay)^2 - axy - a^2y^2 + y^2 + 1$$
$$= (-x - ay)^2 + a(-x - ay)y + y^2 + 1,$$

hence $(-x - ay, y)$ is also a solution. It follows that $-x - ay > 0$.

If we prove that $-x - ay < x$, then we contradict the minimality of x. Indeed, from the symmetry of the equation, we may assume that $x > y$. Then $x^2 > y^2 + 1 = x(-x - ay)$, so $x > -x - ay$. It follows that in this case the equation (6.3.1) is not solvable.

Consider now $a > 0$ and let (x, y) be a solution. Then $xy < 0$ and we may assume for example that $x > 0$ and $y < 0$. Setting $z = -y$, we obtain the equivalent equation $x^2 + (-a)xz + z^2 = -1$, with $x > 0$, $z > 0$, which we examined above. It follows that the equation (6.3.1) is not solvable if $-a \neq -3$, i.e., $a \neq 3$.

It remains to solve the equation when $a = \pm 3$. First, consider the case $a = -3$ and write the equation $x^2 - 3xy + y^2 = -1$ in the following equivalent form $(2x - 3y)^2 - 5y^2 = -4$. This is a special Pell's equation:

$$u^2 - 5v^2 = -4. \tag{6.3.2}$$

Its minimal solution is $(1, 1)$. By the results in Section 4.3.2, it follows that the general solution (u_m, v_m) to (6.3.2) is given by

$$u_m + v_m\sqrt{5} = 2\left(\frac{1 + \sqrt{5}}{2}\right)^m, \quad m = 1, 3, 5, \ldots$$

Since

$$u_m - v_m\sqrt{5} = 2\left(\frac{1 - \sqrt{5}}{2}\right)^m, \quad m = 1, 3, 5, \ldots$$

we obtain

$$u_m = \left(\frac{1 + \sqrt{5}}{2}\right)^m + \left(\frac{1 - \sqrt{5}}{2}\right)^m$$

and

$$v_m = \frac{1}{\sqrt{5}}\left[\left(\frac{1 + \sqrt{5}}{2}\right)^m - \left(\frac{1 - \sqrt{5}}{2}\right)^m\right] = F_m$$

where $m = 1, 3, 5, \ldots$

It follows that $2x - 3F_m = u_m$, hence

$$x = \left(\frac{3}{2\sqrt{5}} + \frac{1}{2}\right)\left(\frac{1+\sqrt{5}}{2}\right)^m - \left(\frac{3}{2\sqrt{5}} - \frac{1}{2}\right)\left(\frac{1-\sqrt{5}}{2}\right)^m$$

$$= \frac{(1+\sqrt{5})^2}{4\sqrt{5}}\left(\frac{1+\sqrt{5}}{2}\right)^m - \frac{(1-\sqrt{5})^2}{2\sqrt{5}}\left(\frac{1-\sqrt{5}}{2}\right)^m$$

$$= \frac{1}{\sqrt{5}}\left[\left(\frac{1+\sqrt{5}}{2}\right)^{m+2} - \left(\frac{1-\sqrt{5}}{2}\right)^{m+2}\right] = F_{m+2}.$$

We obtain the solutions (F_{2n+1}, F_{2n-1}), $n \geq 1$, and by the symmetries $(x, y) \rightarrow (y, x)$ and $(x, y) \rightarrow (-x, -y)$ we find the others.

If $a = 3$, the substitution $y = -z$ transforms the equation into

$$x^2 - 3xz + z^2 = -1.$$

From the above considerations we obtain the solutions

$$(x, z) = (F_{2n+1}, F_{2n-1})$$

and by using the same symmetries we get the four families of solutions given in the Theorem.   □

Remarks. 1) The conclusion in Theorem 6.3.1 can be also obtained by considering the more general equation (see [200] or [25]):

$$x^2 + y^2 + 1 = xyz.$$

The integral solutions (x, y, z) to this equation are given by

$$(-F_{2n-1}, -F_{2n+1}, 3),\ (-F_{2n+1}, -F_{2n-1}, 3),\ (F_{2n-1}, F_{2n+1}, 3),$$
$$(F_{2n+1}, F_{2n-11}, 3),\ (-F_{2n-1}, F_{2n+1}, -3),\ (-F_{2n+1}, F_{2n-1}, -3),$$
$$(F_{2n-1}, -F_{2n+1}, -3),\ (F_{2n+1}, -F_{2n-1}, -3),\ n \geq 1.$$

2) In [150] it is considered the more general equation

$$f_1(x, y) = zf_2(x, y)$$

where $f_1(x, y) = ax^2 + bxy + cy^2 + dx + ey + f$, $f_2(x, y) = pxy + qx + ry + s$ are quadratic forms with integer coefficients and $ac \neq 0$, $a|\gcd(b, d, p, q)$, $c|\gcd(b, e, p, r)$.

3) In the paper [188] is considered the equation

$$x^4 - 6x^2y^2 + 5y^4 = 16F_{n-1}F_{n+1},$$ (6.3.3)

when one of the Fibonacci numbers F_{n-1}, F_{n+1} is prime and another is prime or it is a product of two different prime numbers. There are such Fibonacci numbers, for example $F_5 = 5$ and $F_7 = 13$; $F_{11} = 89$ and $F_{13} = 233$; $F_{17} = 1597$ and $F_{19} = 165580141 = F_{29} = 514229$ and $F_{31} = 1346269 = 557 \cdot 2147$; $F_{41} = 165580141 = 2789 \cdot 59369$ and $F_{43} = 433494437$.

Using the equivalent form $(x^2 - 5y^2)(x^2 - y^2) = 16F_{n-1}F_{n+1}$ to the equation, and the result in Theorem 4.1.1, in the paper [188] is shown that all integral solutions (x, y, n) to (6.3.3) are $(x, y, n) = (\pm L_{6l}, \pm F_{6l}, 6l)$, $l \geq 1$, when $6l - 1$ are prime numbers, F_{6l+1} is a product of two different primes, and L_{6l} is the Lucas number.

6.4 Diophantine Representations of the Sequences Fibonacci, Lucas, and Pell

In this section we will consider some special cases of the Diophantine equation

$$x^2 + axy - y^2 = b$$ (6.4.1)

where a and b are integers and we will show that all nontrivial positive solutions to (6.4.1) are representable by pairs of consecutive terms in the sequences $(F_n)_{n \geq 1}$, $(L_n)_{n \geq 1}$, $(P_n)_{n \geq 1}$. These results are given in [47] but the method used there is different and more complicated. Note that this equation is a special case of (4.8.1).

Theorem 6.4.1. (i) The nontrivial positive integer solutions to the equation

$$x^2 + xy - y^2 = -1$$ (6.4.2)

are given by (F_{2n}, F_{2n+1}), $n \geq 1$.
(ii) The nontrivial positive integer solutions to the equation

$$x^2 + xy - y^2 = 1$$ (6.4.3)

are given by (F_{2n-1}, F_{2n}), $n \geq 1$.

Proof. (i) The equation is equivalent to

$$(2x + y)^2 - 5y^2 = -4.$$

This is a special Pell's equation of the form $u^2 - 5v^2 = -4$ and has solution

$$\frac{1}{2}(u_m + v_m\sqrt{5}) = \left(\frac{1+\sqrt{5}}{2}\right)^m, \quad m = 3, 5, \ldots$$

(see Theorem 4.4.1). It follows that

$$u_m = \left(\frac{1+\sqrt{5}}{2}\right)^m + \left(\frac{1-\sqrt{5}}{2}\right)^m$$

and

$$v_m = \frac{1}{\sqrt{5}}\left[\left(\frac{1+\sqrt{5}}{2}\right)^m - \left(\frac{1-\sqrt{5}}{2}\right)^m\right], \quad m = 3, 5, \ldots$$

Hence $y_m = v_m$ and

$$x_m = \frac{1}{2}(u_m - v_m) = \frac{1}{2}\left[\left(1 - \frac{1}{\sqrt{5}}\right)\left(\frac{1+\sqrt{5}}{2}\right)^m + \left(1 + \frac{1}{\sqrt{5}}\right)\left(\frac{1-\sqrt{5}}{2}\right)^m\right]$$

$$= \frac{1}{\sqrt{5}}\left[\left(\frac{1+\sqrt{5}}{2}\right)^{m-1} - \left(\frac{1-\sqrt{5}}{2}\right)^{m-1}\right], \quad m = 3, 5, \ldots$$

Thus $(x_n, y_n) = (F_{2n}, F_{2n+1})$, $n \geq 1$.

(ii) Similarly, we obtain the equivalent equation

$$(2x + y)^2 - 5y^2 = 4$$

which is a special Pell's equation of the form $u^2 - 5v^2 = 4$ and has solution

$$u_n = \left(\frac{3+\sqrt{5}}{2}\right)^n + \left(\frac{3-\sqrt{5}}{2}\right)^n,$$

$$v_n = \frac{1}{\sqrt{5}}\left[\left(\frac{3+\sqrt{5}}{2}\right)^n - \left(\frac{3-\sqrt{5}}{2}\right)^n\right],$$

where $n \geq 1$ (see (4.3.2)).

It follows that

$$y_n = v_n = \frac{1}{\sqrt{5}}\left[\left(\frac{1+\sqrt{5}}{2}\right)^{2n} - \left(\frac{1-\sqrt{5}}{2}\right)^{2n}\right] = F_{2n}$$

and

$$x_n = \frac{1}{2}(u_n - v_n) = \frac{1}{2}\left[\left(1 - \frac{1}{\sqrt{5}}\right)\left(\frac{1+\sqrt{5}}{2}\right)^{2n} + \left(1 + \frac{1}{\sqrt{5}}\right)\left(\frac{1-\sqrt{5}}{2}\right)^{2n}\right]$$

$$= \frac{1}{\sqrt{5}}\left[\left(\frac{1+\sqrt{5}}{2}\right)^{2n-1} - \left(\frac{1-\sqrt{5}}{2}\right)^{2n-1}\right] = F_{2n-1}.$$

□

Theorem 6.4.2.   *(i) The nontrivial positive integer solutions to the equation*

$$x^2 + xy - y^2 = -5 \tag{6.4.4}$$

are given by (L_{2n-1}, L_{2n}), $n \geq 1$.
(ii) The nontrivial positive integer solutions to the equation

$$x^2 + xy - y^2 = 5 \tag{6.4.5}$$

are given by (L_{2n}, L_{2n+1}), $n \geq 1$.

Proof. Recall that the general term of the Lucas sequence is given by

$$L_m = \left(\frac{1+\sqrt{5}}{2}\right)^m + \left(\frac{1-\sqrt{5}}{2}\right)^m, \quad m \geq 1. \tag{6.4.6}$$

(i) Write the equation in the equivalent form

$$(2x + y)^2 - 5y^2 = -20$$

and let $2x + y = 5u$, $y = v$. We obtain the special Pell's equation $v^2 - 5u^2 = 4$, whose solutions are

$$v_n = \left(\frac{3+\sqrt{5}}{2}\right)^n + \left(\frac{3-\sqrt{5}}{2}\right)^n,$$

$$u_n = \frac{1}{\sqrt{5}}\left[\left(\frac{3+\sqrt{5}}{2}\right)^n - \left(\frac{3-\sqrt{5}}{2}\right)^n\right], \quad n \geq 1.$$

It follows that

$$y_n = v_n = \left(\frac{1+\sqrt{5}}{2}\right)^{2n} + \left(\frac{1-\sqrt{5}}{2}\right)^{2n} = L_{2n}$$

and

$$x_n = \frac{1}{2}(5u_n - v_n) = \left[(\sqrt{5} - 1)\left(\frac{1+\sqrt{5}}{2}\right)^{2n} - (\sqrt{5}+1)\left(\frac{1-\sqrt{5}}{2}\right)^{2n}\right]$$

$$= \left(\frac{1+\sqrt{5}}{2}\right)^{2n-1} + \left(\frac{1-\sqrt{5}}{2}\right)^{2n-1} = L_{2n-1}.$$

(ii) Similarly, the equivalent equation $(2x+y)^2 - 5y^2 = 20$ reduces to $v^2 - 5u^2 = -4$, where $2x+y = 5u$ and $y = v$. We have

$$\frac{1}{2}(v_m + u_m\sqrt{5}) = \left(\frac{1+\sqrt{5}}{2}\right)^m, \quad m = 1, 3, 5, \ldots$$

(see Theorem 4.4.1), hence

$$y_m = v_m = \left(\frac{1+\sqrt{5}}{2}\right)^m + \left(\frac{1-\sqrt{5}}{2}\right)^m = L_m, \quad m = 1, 3, 5, \ldots$$

and

$$x_m = \frac{1}{2}(5u_m - v_m) = \frac{1}{2}\left[(\sqrt{5}-1)\left(\frac{1+\sqrt{5}}{2}\right)^m - (\sqrt{5}+1)\left(\frac{1-\sqrt{5}}{2}\right)^m\right]$$

$$= \left(\frac{1+\sqrt{5}}{2}\right)^{m-1} + \left(\frac{1-\sqrt{5}}{2}\right)^{m-1} = L_{m-1}, \quad m = 1, 3, 5, \ldots$$

$\square$

Theorem 6.4.3. *(i) The nontrivial positive integer solutions to the equation*

$$x^2 + 2xy - y^2 = -1 \tag{6.4.7}$$

are given by (P_{2n}, P_{2n+1}), $n \geq 0$.
(ii) The nontrivial positive integer solutions to the equation

$$x^2 + 2xy - y^2 = 1$$

are given by (P_{2n-1}, P_{2n}), $n \geq 1$.

Proof. The general term of the Pell's sequence is given by

$$P_m = \frac{1}{2\sqrt{2}}[(1 + \sqrt{2})^m - (1 - \sqrt{2})^m], \quad m \geq 1. \tag{6.4.8}$$

(i) Write the equation in the equivalent form $(x+y)^2 - 2y^2 = -1$. This is a negative Pell's equation of the form $u^2 - 2v^2 = -1$, whose solutions are given by

$$u_n = \frac{1}{2}[(1+\sqrt{2})^{2n+1} + (1-\sqrt{2})^{2n+1}]$$

and

$$v_n = \frac{1}{2\sqrt{2}}[(1+\sqrt{2})^{2n+1} - (1-\sqrt{2})^{2n+1}], \quad n \geq 0.$$

It follows that

$$y_n = v_n = P_{2n+1}$$

and

$$x_n - u_n - v_n = \frac{1}{2}\left[\left(1 - \frac{1}{\sqrt{2}}\right)(1+\sqrt{2})^{2n+1} + \left(1 + \frac{1}{\sqrt{2}}\right)(1-\sqrt{2})^{2n+1}\right]$$

$$= \frac{1}{2\sqrt{2}}[(1+\sqrt{2})^{2n} - (1-\sqrt{2})^{2n}] = P_{2n}.$$

(ii) We obtain the Pell's equation $(x+y)^2 - 2y^2 = 1$, whose solutions are

$$x_n + y_n = \frac{1}{2}[(1+\sqrt{2})^{2n} + (1-\sqrt{2})^{2n}],$$

$$y_n = \frac{1}{2\sqrt{2}}[(1+\sqrt{2})^{2n} - (1-\sqrt{2})^{2n}], \quad n \geq 1.$$

It follows that $y_n = P_{2n}$ and

$$x_n = \frac{1}{2}\left[\left(1 - \frac{1}{\sqrt{2}}\right)(1+\sqrt{2})^{2n} + \left(1 + \frac{1}{\sqrt{2}}\right)(1-\sqrt{2})^{2n}\right]$$

$$= \frac{1}{2\sqrt{2}}[(1+\sqrt{2})^{2n-1} - (1-\sqrt{2})^{2n-1}] = P_{2n-1}.$$

□

The results in Theorems 6.4.1–6.4.3 can be summarized in the following Theorem proven by the infinite descent method in [47].

Theorem 6.4.4. *Let a be a positive integer and let $(\alpha_n)_{n\geq 1}$ be the sequence defined recursively by*

$$\begin{cases} \alpha_1 = 1, \ \alpha_2 = a \\ \alpha_{n+1} = a\alpha_n + \alpha_{n-1}, \ n \geq 2. \end{cases} \tag{6.4.9}$$

Then all positive integer solutions to the equation

$$|x^2 + axy - y^2| = 1 \tag{6.4.10}$$

are given by (α_n, α_{n+1}), $n \geq 1$.

Proof. The general term of the sequence $(\alpha_n)_{n\geq 1}$ in (6.4.9) is given by

$$\alpha_n = \frac{1}{\sqrt{a^2+4}} \left[\left(\frac{a+\sqrt{a^2+4}}{2} \right)^n - \left(\frac{a-\sqrt{a^2+4}}{2} \right)^n \right], \quad n \geq 1. \tag{6.4.11}$$

The equation $x^2 + axy - y^2 = -1$ is equivalent to $(2x+ay)^2 - (a^2+4)y^2 = -4$, which is a special Pell's equation of the form $u^2 - (a^2+4)v^2 = -4$. From Theorem 4.4.1 it follows that

$$\frac{1}{2}(u_m + v_m \sqrt{a^2+4}) = \left(\frac{a+\sqrt{a^2+4}}{2} \right)^m, \quad m = 1, 3, 5, \ldots$$

Hence

$$u_m = \left(\frac{a+\sqrt{a^2+4}}{2} \right)^m + \left(\frac{a-\sqrt{a^2+4}}{2} \right)^m$$

and

$$v_m = \frac{1}{\sqrt{a^2+4}} \left[\left(\frac{a+\sqrt{a^2+4}}{2} \right)^m - \left(\frac{a-\sqrt{a^2+4}}{2} \right)^m \right].$$

Therefore $y_m = v_m = \alpha_m$, $m = 1, 3, 5, \ldots$, and

$$x_m = \frac{1}{2}(u_m - av_m)$$

$$= \frac{1}{2} \left[\left(1 - \frac{a}{\sqrt{a^2+4}} \right) \left(\frac{a+\sqrt{a^2+4}}{2} \right)^m + \left(1 + \frac{a}{\sqrt{a^2+4}} \right) \left(\frac{a-\sqrt{a^2+4}}{2} \right)^m \right]$$

$$= \frac{1}{\sqrt{a^2+4}} \left[\left(\frac{a+\sqrt{a^2+4}}{2} \right)^{m-1} - \left(\frac{a-\sqrt{a^2+4}}{2} \right)^{m-1} \right] = \alpha_{m-1}.$$

The equation $x^2 + axy - y^2 = 1$ is equivalent to $(2x+ay)^2 - (a^2+4)y^2 = 4$. From Theorem 4.4.1 it follows that the general solution to the equation $u^2 - (a^2+4)v^2 = 4$ is given by

$$\frac{1}{2}(u_n + v_n\sqrt{a^2 + 4}) = \left(\frac{a^2 + 2 + a\sqrt{a^2 + 4}}{2}\right)^n = \left(\frac{a + \sqrt{a^2 + 4}}{2}\right)^{2n}.$$

We obtain

$$u_n = \left(\frac{a + \sqrt{a^2 + 4}}{2}\right)^{2n} + \left(\frac{a - \sqrt{a^2 + 4}}{2}\right)^{2n},$$

$$v_n = \frac{1}{\sqrt{a^2 + 4}}\left[\left(\frac{a + \sqrt{a^2 + 4}}{2}\right)^{2n} - \left(\frac{a - \sqrt{a^2 + 4}}{2}\right)^{2n}\right].$$

Hence $y_n = v_n = \alpha_{2n}$ and

$$x_n = \frac{1}{2}(u_n - av_n)$$

$$= \frac{1}{2}\left[\left(1 - \frac{a}{\sqrt{a^2 + 4}}\right)\left(\frac{a + \sqrt{a^2 + 4}}{2}\right)^{2n} + \left(1 + \frac{a}{\sqrt{a^2 + 4}}\right)\left(\frac{a - \sqrt{a^2 + 4}}{2}\right)^{2n}\right]$$

$$= \frac{1}{\sqrt{a^2 + 4}}\left[\left(\frac{a + \sqrt{a^2 + 4}}{2}\right)^{2n-1} - \left(\frac{a - \sqrt{a^2 + 4}}{2}\right)^{2n-1}\right] = \alpha_{2n-1}.$$

□

Remarks. 1) Theorem 6.4.4 characterizes the pairs of consecutive terms of the sequence $(\alpha_n)_{n\geq 1}$ defined by the linear recurrence (6.4.9).

2) The set consisting of α_{2k}, $k \geq 1$, is included in the set of positive values of the polynomial

$$P_1(x, y) = x[1 - (x^2 + axy - y^2 - 1)^2]$$

and the set consisting of α_{2k+1}, $k \geq 0$, is included in the set of positive values of the polynomial

$$P_2(x, y) = x[1 - (x^2 + axy - y^2 + 1)^2].$$

3) The result in Theorem 6.4.4 also appears in the paper [134].

6.5   Diophantine Representations of Generalized Lucas Sequences

We define the *generalized Lucas sequence* $(\gamma_n)_{n\geq 0}$, $\gamma_n = \gamma_n(a,b)$, with parameters $a, b \in \mathbb{Z}^*$ by

$$\gamma_{n+1} = f(a,b)\gamma_n - \gamma_{n-1}, \quad n \geq 1 \tag{6.5.1}$$

where $f : \mathbb{Z}^* \times \mathbb{Z}^* \to \mathbb{Z}$ is a given function and $\gamma_0 = \gamma_0(a,b)$, $\gamma_1 = \gamma_1(a,b)$ are given integers.

The following theorems generalize all results in Section 6.4.

Theorem 6.5.1. *Let a, b be nonzero integers such that $b \neq 1$ and $a^2 - 4b > 0$ is a nonsquare. All integral solutions to the equation*

$$x^2 + axy + by^2 = 1 \tag{6.5.2}$$

are given by $(\alpha_n, \beta_n)_{n\geq 1}$, $(-\alpha_n, -\beta_n)_{n\geq 1}$, where $(\alpha_n)_{n\geq 1}$, $(\beta_n)_{n\geq 1}$ are the generalized Lucas sequences defined by

$$\alpha_{n+1} = u_0\alpha_n - \alpha_{n-1}, \ \alpha_0 = 2, \ \alpha_1 = u_0 \text{ and}$$

$$\beta_{n+1} = u_0\beta_n - \beta_{n-1}, \ \beta_0 = 1, \ \beta_1 = \frac{1}{2}(u_0 - av_0). \tag{6.5.3}$$

Here $u_0 = u_0(a,b)$, $v_0 = v_0(a,b)$ are the minimal solutions to the special Pell's equation

$$u^2 - (a^2 - 4b)v^2 = 4. \tag{6.5.4}$$

Proof. The general terms of the sequences $(\alpha_n)_{n\geq 0}$ and $(\beta_n)_{n\geq 0}$ are given by

$$\alpha_n = \left(\frac{u_0 + v_0\sqrt{a^2 - 4b}}{2}\right)^n + \left(\frac{u_0 - v_0\sqrt{a^2 - 4b}}{2}\right)^n$$

and

$$\beta_n = \frac{1}{2}\left[\left(1 - \frac{a}{\sqrt{a^2 - 4b}}\right)\left(\frac{u_0 + v_0\sqrt{a^2 - 4b}}{2}\right)^n \right.$$
$$\left. + \left(1 + \frac{a}{\sqrt{a^2 - 4b}}\right)\left(\frac{u_0 - v_0\sqrt{a^2 - 4b}}{2}\right)^n\right].$$

The equation (6.5.2) is equivalent to $(2x + ay)^2 - (a^2 - 4b)y^2 = 4$, i.e., to the general Pell's equation (6.5.3). From Theorem 4.4.1 its general solution is given by

$$\frac{1}{2}(u_n + v_n\sqrt{a^2 - 4b}) = \left(\frac{u_0 + v_0\sqrt{a^2 - 4b}}{2}\right)^n, \quad n \geq 1.$$

It follows that

$$u_n = \left(\frac{u_0 + v_0\sqrt{a^2 - 4b}}{2}\right)^n + \left(\frac{u_0 - v_0\sqrt{a^2 - 4b}}{2}\right)^n$$

and

$$v_n = \frac{1}{\sqrt{a^2 - 4b}}\left[\left(\frac{u_0 + v_0\sqrt{a^2 - 4b}}{2}\right)^n - \left(\frac{u_0 - v_0\sqrt{a^2 - 4b}}{2}\right)^n\right], \quad n \geq 1.$$

Thus $y_n = v_n = \alpha_n$ and $x_n = \frac{1}{2}(u_n - av_n) = \beta_n$.   □

Remark. Theorems 6.5.1 and 6.5.2 give an useful method for solving the Diophantine equations of degree three in four variables

$$x^2 + uxy + vy^2 = \pm 1.$$

Indeed, setting $u = a$, $v = b$, with $a, b \in \mathbb{Z}$, the above equations are equivalent to

$$(2x + ay)^2 - (a^2 - 4b)y^2 = \pm 4.$$

If $a^2 - 4b < 0$, there are at most finitely many solutions.

If $a^2 - 4b = 0$, the equations reduce to $(2x + ay)^2 = \pm 4$, and for a even we obtain solutions $\left(-\frac{ka}{2} \pm 1, k\right), k \in \mathbb{Z}$.

If $a^2 - 4b > 0$ is a perfect square, there are at most finitely many solutions.

If $a^2 - 4b > 0$ is not a square, then all solutions to the equation $x^2 + uxy + vy^2 = 1$ are given by $(x, y, u, v) = (\pm\alpha_m, \pm\beta_m, a, b)$, $m \geq 1$, where (α_m), (β_m) are the generalized Lucas sequences defined in Theorem 6.5.1.

All solutions to the equation $x^2 + uxy + vy^2 = -1$ are given by

$$(x, y, u, v) = (\pm\alpha_{2n+1}, \pm\beta_{2n+1}, a, b), \quad n \geq 0,$$

where (α_m), (β_m) are defined in Theorem 6.5.2.

For some particular values of a and b the generalized Lucas sequences $(\alpha_n)_{n \geq 0}$ and $(\beta_n)_{n \geq 0}$ defined by (6.5.3) coincide with some classical sequences. In the following table we will give a few such situations (see also Section 6.4).

a	b	Equation (6.5.2)	Solutions
1	-1	$x^2 + xy - y^2 = 1$	$(F_{2n-1}, F_{2n}), (-F_{2n-1}, -F_{2n})$
-1	-1	$x^2 - xy - y^2 = 1$	$(F_{2n+1}, F_{2n}), (-F_{2n+1}, -F_{2n})$
2	-1	$x^2 + 2xy - y^2 = 1$	$(P_{2n-1}, P_{2n}), (-P_{2n-1}, -P_{2n})$
-2	-1	$x^2 - 2xy - y^2 = 1$	$(P_{2n+1}, P_{2n}), (-P_{2n+1}, -P_{2n})$

Theorem 6.5.2. *Let a, b be nonzero integers such that $b \neq 1$ and $a^2 - 4b > 0$ is a nonsquare. Assume that the special Pell's equation*

$$s^2 - (a^2 - 4b)t^2 = -4 \tag{6.5.5}$$

is solvable and its minimal solution is (s_0, t_0), $s_0 = s_0(a, b)$, $t_0 = t_0(a, b)$. Then all integral solutions to the equation

$$x^2 + axy + by^2 = -1 \tag{6.5.6}$$

are given $(\alpha_{2n+1}, \beta_{2n+1})$, $(-\alpha_{2n+1}, -\beta_{2n+1})$, $n \geq 1$, where $(\alpha_m)_{m \geq 0}$, $(\beta_m)_{m \geq 0}$ are the generalized Lucas sequences defined by

$$\alpha_{m+1} = s_0 \alpha_m - \alpha_{m-1}, \ \alpha_0 = 2, \ \alpha_1 = s_0 \text{ and}$$

$$\beta_{m+1} = s_0 \beta_m - \beta_{m-1}, \ \beta_0 = 1, \ \beta_1 = \frac{1}{2}(s_0 - at_0). \tag{6.5.7}$$

Proof. We proceed like in the previous theorem and take into account the results in Theorem 4.4.1 concerning the general solution to the equation (6.5.5). □

In some special cases, the generalized Lucas sequences defined by (6.5.7) yield to solutions involving well-known sequences. We will illustrate this by presenting the following table (see also Section 6.5).

a	b	Equation (6.5.6)	Solutions
1	-1	$x^2 + xy - y^2 = -1$	$(F_{2n}, F_{2n+1}), (-F_{2n}, -F_{2n+1})$
-1	-1	$x^2 - xy - y^2 = -1$	$(F_{2n}, F_{2n-1}), (-F_{2n}, -F_{2n-1})$
2	-1	$x^2 + 2xy - y^2 = -1$	$(P_{2n}, P_{2n+1}), (-P_{2n}, -P_{2n+1})$
-2	-1	$x^2 - 2xy - y^2 = -1$	$(P_{2n}, P_{2n-1}), (-P_{2n}, -P_{2n-1})$
5	5	$x^2 + 5xy + 5y^2 = -1$	$(L_{2n}, -F_{2n+1}), (-L_{2n}, F_{2n+1})$

Remarks. 1) The solvability condition for the general Pell's equation (6.5.5) in Theorem 6.5.2 is necessary. Indeed, for example, for $a = 5, b = 1$ the equation

$$s^2 - 21t^2 = -4$$

is not solvable (the left-hand side is congruent to 0 or 1 (mod 3)). The corresponding equation (6.5.6):

$$x^2 + 5xy - y^2 = -1$$

is also not solvable.

2) The special case $b = -1$ is studied in [134]. Particular Diophantine representations for the Fibonacci and Lucas sequences are given in [96] and [97]. We also mention the connection with the general Pell's equation given in [61]. A particular definition for generalized Lucas sequences appears in [91].

An interesting special case for the equation (6.5.2) is $a = b$. We obtain the Diophantine equation

$$x^2 + axy + ay^2 = 1. \tag{6.5.8}$$

The general Pell's equation (6.5.4) becomes $u^2 - (a^2 - 4a)v^2 = 4$, whose minimal solution is $(u_0, v_0) = (a - 2, 1)$.

With the notations in Theorem 6.5.1 the generalized Lucas sequences $(\alpha_n)_{n \geq 0}, (\beta_n)_{n \geq 0}$ are given by

$$\alpha_{n+1} = (a - 2)\alpha_n - \alpha_{n-1}, \quad \alpha_0 = 2, \quad \alpha_1 = a - 2$$
$$\beta_{n+1} = (a - 2)\beta_n - \beta_{n-1}, \quad \beta_0 = 1, \quad \beta_1 = -1.$$

From Theorem 6.5.1 we obtain the following Corollary:

Corollary 6.5.3. *The equation (6.5.8) is always solvable and all of its solutions are given by* $(\alpha_n, \beta_n)_{n \geq 0}$.

Next we study when the solutions to the equation (6.5.8) are linear combinations over $\mathbb{Q}$ of the classical Fibonacci and Lucas sequences. The results are obtained in the paper [20].

For other results we refer to the papers [61, 91, 96–98] and [134]. Also, the problem is connected to the Y.V. Matiasevich and J. Robertson way to solve the Hilbert's Tenth Problem, and it has applications to the problem of singlefold Diophantine representation of recursively enumerable sets. In the recent paper [102] the equations $x^2 - kxy + y^2 = 1, x^2 - kxy - y^2 = 1$ are solved in terms of generalized Fibonacci and Lucas numbers. Let us mention that in the paper [83] is defined the Hankel matrices involving the Pell, Pell-Lucas and modified Pell sequences, is computed their Frobenius norm, and it is investigated some spectral properties of them.

Recall the Binet's formulas for F_n and L_n:

$$F_n = \frac{1}{\sqrt{5}} \left[\left(\frac{1+\sqrt{5}}{2} \right)^n - \left(\frac{1-\sqrt{5}}{2} \right)^n \right],$$

$$L_n = \left(\frac{1+\sqrt{5}}{2} \right)^n + \left(\frac{1-\sqrt{5}}{2} \right)^n.$$

These formulas can be extended to negative integers n in a natural way. We have $F_{-n} = (-1)^{n-1}F_n$ and $L_{-n} = (-1)^n L_n$, for all n.

Theorem 6.5.4. *The solutions to the positive equation (6.5.8) are linear combinations with rational coefficients of at most two Fibonacci and Lucas numbers if and only if $a = a_n = \pm L_{2n} + 2$, $n \geq 1$.*

For each n, all of its integer solutions (x_k, y_k) are given by

$$
\begin{cases}
x_k = \dfrac{\varepsilon_k}{2} L_{2kn} \mp \dfrac{a_n}{2F_{2n}} F_{2kn} \\
y_k = \pm \dfrac{1}{F_{2n}} F_{2kn},
\end{cases}
\tag{6.5.9}
$$

where $k \geq 1$, signs $+$ and $-$ depend on k and correspond, while $\varepsilon_k = \pm 1$.

Proof. The equation $x^2 + axy + ay^2 = 1$ is equivalent to the positive special Pell's equation

$$(2x + ay)^2 - (a^2 - 4a)y^2 = 4. \tag{6.5.10}$$

From formula (4.4.6) it follows that

$$2x_n + ay_m = \varepsilon_m \left[\left(\frac{u_1 + v_1\sqrt{D}}{2} \right)^m + \left(\frac{u_1 - v_1\sqrt{D}}{2} \right)^m \right]$$

and

$$y_m = \frac{\varepsilon_m}{\sqrt{D}} \left[\left(\frac{u_1 + v_1\sqrt{D}}{2} \right)^m - \left(\frac{u_1 - v_1\sqrt{D}}{2} \right)^m \right],$$

where $m \in \mathbb{Z}$, $\varepsilon_m = \pm 1$, $D = a^2 - 4a$, and (u_1, v_1) is the minimal positive solution to $u^2 - Dv^2 = 4$. we have $(u_1, v_1) = (a - 2, 1)$, and combining the above relations it follows

$$x_m = \frac{\varepsilon_m}{2} \left[\left(1 - \frac{a}{\sqrt{a^2 - 4a}}\right) \left(\frac{a - 2 + \sqrt{a^2 - 4a}}{2}\right)^m \right.$$

$$\left. + \left(1 + \frac{a}{\sqrt{a^2 - 4a}}\right) \left(\frac{a - 2 - \sqrt{a^2 - 4a}}{2}\right)^m \right] \tag{6.5.11}$$

and

$$y_m = \frac{\varepsilon_m}{\sqrt{a^2 - 4a}} \left[\left(\frac{a - 2 + \sqrt{a^2 - 4a}}{2}\right)^m - \left(\frac{a - 2 - \sqrt{a^2 - 4a}}{2}\right)^m \right]. \tag{6.5.12}$$

Taking into account Binet's formulas, solution (x_m, y_m) is representable in terms of F_m and L_m only if $a^2 - 4a = 5s^2$, for some positive integer s. This is equivalent to the special Pell's equation

$$(a - 2)^2 - 5s^2 = 4, \tag{6.5.13}$$

whose minimal solution is $(a_1 - 2, s_1) = (3, 1)$. The general integer solution to (6.5.13) is

$$a_n - 2 = \varepsilon_n \left[\left(\frac{3 + \sqrt{5}}{2}\right)^n + \left(\frac{3 - \sqrt{5}}{2}\right)^n \right] = \varepsilon_n L_{2n},$$

and

$$s_n = \frac{\varepsilon_n}{\sqrt{5}} \left[\left(\frac{3 + \sqrt{5}}{2}\right)^n - \left(\frac{3 - \sqrt{5}}{2}\right)^n \right] = \varepsilon_n F_{2n},$$

where n is an integer and $\varepsilon_n = \pm 1$.

From $(2x + ay)^2 - (a^2 - 4a)y^2 = 4$ we find $(2x + a_n y)^2 - 5(s_n y)^2 = 4$, with integer solution (x_m, y_m) given by

$$2x_m + a_n y_m = \varepsilon_{2m} L_{2m} \quad \text{and} \quad s_n y_m = \pm F_{2m}.$$

Hence

$$x_m = \frac{1}{2} \left[\varepsilon_{2m} L_{2m} \mp a_n \frac{F_{2m}}{F_{2n}} \right], \quad y_m = \pm \frac{F_{2m}}{F_{2n}}, \tag{6.5.14}$$

where signs $+$ and $-$ correspond, and $\varepsilon_{2m} = \pm 1$.

Taking into account that F_{2n} divides F_{2m} if and only if n divides m (see [21, p. 180] and [90, p. 39]), it is necessary that $m = kn$, for some positive integer k. Formula (6.5.14) becomes (6.5.9).

A parity argument shows that in the equation

$$(x + ay)^2 - (a^2 - 4a)y^2 = 4,$$

X is even, so x_k in (6.5.9) is always an integer.   □

The following two tables give the integer solutions to equation (6.5.8) at level k, including the trivial solution obtained for $k = 0$.

n	$a_n = L_{2n} + 2$	Equation (6.5.8)	Solutions
1	5	$x^2 + 5xy + 5y^2 = 1$	$x = \frac{\varepsilon_k}{2}L_{2k} \mp \frac{5}{2}F_{2k},\ y = \pm F_{2k}$
2	9	$x^2 + 9xy + 9y^2 = 1$	$x = \frac{\varepsilon_k}{2}L_{4k} \mp \frac{5}{6}F_{4k},\ y = \pm\frac{1}{3}F_{4k}$
3	20	$x^2 + 20xy + 20y^2 = 1$	$x = \frac{\varepsilon_k}{2}L_{6k} \mp \frac{5}{4}F_{6k},\ y = \pm\frac{1}{8}F_{6k}$
4	49	$x^2 + 49xy + 49y^2 = 1$	$x = \frac{\varepsilon_k}{2}L_{8k} \mp \frac{7}{6}F_{8k},\ y = \pm\frac{1}{21}F_{8k}$
5	125	$x^2 + 125xy + 125y^2 = 1$	$x = \frac{\varepsilon_k}{2}L_{10k} \mp \frac{25}{22}F_{10k},\ y = \pm\frac{1}{55}F_{10k}$
6	324	$x^2 + 324xy + 324y^2 = 1$	$x = \frac{\varepsilon_k}{2}L_{12k} \mp \frac{9}{8}F_{12k},\ y = \pm\frac{1}{144}F_{12k}$

n	$a_n = -L_{2n} + 2$	Equation (6.5.8)	Solutions
1	−1	$x^2 - xy - y^2 = 1$	$x = \frac{\varepsilon_k}{2}L_{2k} \pm \frac{1}{2}F_{2k},\ y = \pm F_{2k}$
2	−5	$x^2 - 5xy - 5y^2 = 1$	$x = \frac{\varepsilon_k}{2}L_{4k} \pm \frac{5}{6}F_{4k},\ y = \pm\frac{1}{3}F_{4k}$
3	−16	$x^2 - 16xy - 16y^2 = 1$	$x = \frac{\varepsilon_k}{2}L_{6k} \pm F_{6k},\ y = \pm\frac{1}{8}F_{6k}$
4	−45	$x^2 - 45xy - 45y^2 = 1$	$x = \frac{\varepsilon_k}{2}L_{8k} \pm \frac{15}{14}F_{8k},\ y = \pm\frac{1}{21}F_{8k}$
5	−121	$x^2 - 121xy - 121y^2 = 1$	$x = \frac{\varepsilon_k}{2}L_{10k} \mp \frac{11}{10}F_{10k},\ y = \pm\frac{1}{55}F_{10k}$
6	−320	$x^2 - 320xy - 320y^2 = 1$	$x = \frac{\varepsilon_k}{2}L_{12k} \mp \frac{10}{9}F_{12k},\ y = \pm\frac{1}{144}F_{12k}$

Next we will consider the "negative" equation of the type (6.5.8):

$$x^2 + axy + ay^2 = -1. \tag{6.5.15}$$

Unlike the result in Theorem 6.5.4, there are only two values of a for which the corresponding property holds.

Theorem 6.5.5. *The solutions to the negative equation (6.5.15) are linear combinations with rational coefficients of at most two Fibonacci and Lucas numbers if and only if $a = -1$ or $a = 5$.*
If $a = -1$, all of its integer solutions (x_m, y_m) are given by

$$x_m = \frac{\varepsilon_m}{2}L_{2m+1} \pm \frac{1}{2}F_{2m+1}, \quad y_m = \pm F_{2m+1}, \quad m \geq 0. \tag{6.5.16}$$

If $a = 5$, all integer solutions (x_m, y_m) are

$$x_m = \frac{\varepsilon_m}{2}L_{2m+1} \mp 5F_{2m+1}, \quad y_m = \pm F_{2m+1}, \quad m \geq 0. \tag{6.5.17}$$

The signs $+$ and $-$ depend on m and correspond, while $\varepsilon_m = \pm 1$.

Proof. As in the proof of Theorem 6.5.4 the equation is equivalent to

$$(2x + ay)^2 - (a^2 - 4a)y^2 = -4.$$

Suppose that this negative special Pell's equation is solvable. Its solution (x_m, y_m) is representable in terms of Fibonacci and Lucas numbers as a linear combination with rational coefficients only if $a^2 - 4a = 5s^2$. As in the proof of Theorem 6.5.4 we obtain $a_n = \pm L_{2n} + 2$ and $s_n = \pm F_{2n}$, $n \geq 1$.

The equation $(2x + ay)^2 - (a^2 - 4a)y^2 = -4$ becomes

$$(2x + ay)^2 - 5(s_n y)^2 = -4,$$

whose integer solutions are $2x_m + ay_m = \varepsilon_m L_{2m+1}$ and $s_n y_m = \pm F_{2m+1}$. It follows that

$$y_m = \pm \frac{F_{2m+1}}{F_{2n}}, \quad m \geq 1.$$

If $n \geq 2$, then $F_{2n} \geq 2$, and since $2n$ does not divide $2m + 1$, it follows that F_{2n} does not divide F_{2m+1} (see [21, pp. 180] and [90, pp. 39]), hence y_m is not an integer.

Thus $n = 1$ and so $a = \pm L_2 + 2$, i.e., $a = -1$ or $a = 5$.

For $a = -1$, it follows $y_m = \pm F_{2m+1}$ and $2x_m - y_m = \varepsilon_m L_{2m+1}$, and we obtain solutions (6.5.16).

If $a = 5$, then $y_m = \pm F_{2m+1}$ and $2x_m + 5y_m = \varepsilon_m L_{2m+1}$, yielding the solutions (6.5.17). $\square$

Remark. On the other hand, it is more or less known Zeckendorf's theorem in [230], which states that every positive integer can be represented uniquely as the sum of one or more distinct Fibonacci numbers in such a way that the sum does not include two consecutive Fibonacci numbers. Such a sum is called *Zeckendorf representation* and it is related to the Fibonacci coding of a positive integer. Our results are completely different, because the number of terms is reduced to at most two, and the sum in the representation of solutions is a linear combination with rational coefficients.

Chapter 7
Other Applications

7.1 When Are $an + b$ and $cn + d$ Simultaneously Perfect Squares?

In [122] and [123] it is proven that there are infinitely many positive integers n such that $2n + 1$ and $3n + 1$ are both perfect squares. The proof relies on the theory of general Pell's equations.

In what follows we will present an extension of this result, based on our papers [13] and [14]. The main result is also cited in [41, Problem 1.13]. Recall that in Theorem 4.5.2 we proved that if a, c are relatively prime positive integers, not both perfect squares, and if b, d are integers, then the Diophantine equation

$$ax^2 - cy^2 = ad - bc$$

is solvable if and only if $an + b$ and $cn + d$ are simultaneously perfect squares for some positive integer n. In this case, the number of such n's is infinite. If $(x_m, y_m)_{m \geq 0}$ are solutions to $ax^2 - by^2 = ad - bc$ (see Theorem 4.5.1), then for all n_m m $m \geq 0$, where

$$n_m = \frac{y_m^2 - b}{a} = \frac{x_m^2 - d}{c},$$

$an_m + b$ and $cn_m + d$ are simultaneously perfect squares (see Theorem 4.5.2).

From the previous formulas, we see that the least positive n_0 for which $an_0 + b$ and $cn_0 + d$ are simultaneously perfect squares is

$$n_0 = \frac{x_0^2 - d}{c} = \frac{y_0^2 - b}{a},$$

where (x_0, y_0) is the minimal solution to the equation $ax^2 - cy^2 = ad - bc$.

The main result in this section is the following:

© Springer Science+Business Media New York 2015
T. Andreescu, D. Andrica, *Quadratic Diophantine Equations*,
Developments in Mathematics 40, DOI 10.1007/978-0-387-54109-9_7

Theorem 7.1.1. *Let a, c be relatively prime positive integers, not simultaneously perfect squares and let b, d be integers. Each of the following conditions is sufficient for the numbers $an + b$ and $cn + d$ to be both perfect squares for infinitely many positive integers n:*

1) b and d are perfect squares;
2) $a + b$ and $c + d$ are perfect squares;
3) $\dfrac{a}{c} = \dfrac{b-1}{d-1}.$

Proof. Conditions 1) and 2) state that $an + b$ and $cn + d$ are simultaneously perfect squares for $n = 0$ and $n = 1$, respectively. From Theorem 4.5.2 it follows that they have this property for infinitely many positive integers n.

Condition 3) is equivalent to $a - c = ad - bc$, in which case the equation $ax^2 - by^2 = ad - bc$ has solution $(1, 1)$ and the conclusion follows from the same Theorem 4.5.2.   □

Applications

1) The numbers $2n + 3$ and $5n + 6$ are both perfect squares for infinitely many positive integers n. Indeed, the equation $2x^2 - 5y^2 = -3$ has solution $(1, 1)$ and the result follows from Theorem 4.5.2.
2) If k is an arbitrary positive integer different from 3, then n and $(k^2 - 4)n - 1$ cannot be simultaneously perfect squares. Indeed, in Section 3.6 we saw that the negative Pell's equation $x^2 - (k^2 - 4)y^2 = -1$ is not solvable (see also [199]) and the conclusion follows from Theorem 4.5.2.
3) If p and q are relatively prime positive integers and pq is not a perfect square, then $pn + 1$ and $qn + 1$ are simultaneously perfect squares for infinitely many positive integers n. This property follows from Theorem 7.1.1.1). For $p = 2$ and $q = 3$ we obtain the result in [122]. For $p = 3$ and $q = 4$ we obtain Problem 8 in [25, p. 83].

   If $p = 1$ and $q = 3$ we obtain the first part of the result in [40]. The second part shows that if $n_1 < n_2 < \cdots < n_k < \ldots$ are all positive integers satisfying the above property, then $n_k n_{k+1} + 1$ is also a perfect square, $k = 1, 2, \ldots$. Indeed, the equation is $3x^2 - y^2 = 2$, which is equivalent to the Pell's equation $u^2 - 3v^2 = 1$, where $u = \dfrac{1}{2}(3x - y)$ and $v = \dfrac{1}{2}(y - x)$. The general solution is $(u_k, v_k)_{k \geq 1}$, where $u_k + v_k \sqrt{3} = (2 + \sqrt{3})^k$, $k \geq 1$, hence

$$n_k = x_k^2 - 1 = (u_k + v_k)^2 - 1 = \frac{1}{6}[(2 + \sqrt{3})^{2k+1} + (2 - \sqrt{3})^{2k+1} - 4].$$

   We have

$$n_k n_{k+1} + 1 = \left\{ \frac{1}{6}[(2 + \sqrt{3})^{2k+2} + (2 - \sqrt{3})^{2k+2} - 8] \right\}^2, \quad k \geq 1.$$

4) For any nonzero integers k and l, the numbers $(k^2 + 1)n + 2l$ and $2kn + l^2 + 1$ are simultaneously perfect squares for infinitely many positive integers n. This follows from Theorem 7.1.1.2).

5) The following application appeared in [11] (see also [25, p. 82]). The smallest positive integer m such that $19m + 1$ and $95m + 1$ are both perfect squares is 134232. Indeed, setting $19m = n$ we are looking for the smallest $n \equiv 0$ (mod 19) such that $n + 1$ and $5n + 1$ are simultaneously perfect square. In this case, the equation is $x^2 - 5y^2 = -4$, whose general solution is given by

$$\frac{1}{2}(x_k + y_k\sqrt{5}) = \left(\frac{1 + \sqrt{5}}{2}\right)^k, \quad k = 1, 3, 5, \ldots$$

(see also Section 4.3.2). It follows that $y_k = F_{2k-1}$, $k = 1, 2, \ldots$, and $n_k = F_{2k-1}^2 - 1$. The smallest k for which $n_k \equiv 0$ (mod 19) is $k = 9$, hence the desired integers is $m = \dfrac{1}{19}n_9 = 134232$.

7.2 Triangular Numbers

Let $T_n = \dfrac{n(n+1)}{2}$ denote the nth *triangular number*. In this section we will present several situations when some properties related to these numbers reduce to solving Pell-type equations.

7.2.1 Triangular Numbers with Special Properties

There are infinitely many positive integers n for which T_n is a perfect square. Indeed, if T_n is a perfect square, then so is $T_{4n(n+1)}$, because $\dfrac{n(n+1)}{2} = k^2$ implies

$$T_{4n(n+1)} = T_{8k^2} = 4k^2(8k^2 + 1) = 4k^2(4n^2 + 4n + 1) = [2k(2n + 1)]^2.$$

Taking into account that $T_1 = 1^2$, by the above procedure, we generate a sequence of perfect square triangular numbers. A formula for such integers n has already been given in (5.4.8).

It is natural to ask what are all triangular numbers that are perfect squares. We have [15]:

Theorem 7.2.1. *The triangular number T_x is a perfect square if and only if*

$$x = \begin{cases} 2P_m^2, & m \text{ even} \\ \left[\dfrac{(1+\sqrt{2})^m + (1-\sqrt{2})^m}{2} \right]^2, & m \text{ odd} \end{cases} \qquad (7.2.1)$$

where $(P_m)_{m \geq 0}$ is the Pell's sequence.

Proof. The equation $T_x = y^2$ is equivalent to $(2x+1)^2 - 8y^2 = 1$. The Pell's equation $u^2 - 8v^2 = 1$ has solutions

$$u_m = \frac{1}{2}[(1+\sqrt{2})^{2m} + (1-\sqrt{2})^{2m}]$$

and

$$v_m = \frac{1}{2\sqrt{2}}[(1+\sqrt{2})^{2m} - (1-\sqrt{2})^{2m}] = P_{2m}$$

hence the conclusion.   □

Remarks. 1) Every other x satisfying $T_x = y^2$ is a perfect square.
2) Every y satisfying $T_x = y^2$ is a Pell number.
3) The equation $T_x = (T_y)^2$ is more difficult. It has only solutions $(1, 1)$ and $(8, 3)$. A complicated proof was given by W. Ljunggren (see [150] for details).
4) Some extensions of the result in Theorem 7.2.1 are given in the paper [223].

Theorem 7.2.2. *If k is a positive integer that is not a perfect square, then the equation*

$$kT_x = T_y \qquad (7.2.2)$$

has infinitely many solutions in positive integers.

Proof. Equation (7.2.2) is equivalent to $(2y+1)^2 - k(2x+1)^2 = 1 - k$. Let (u_1, v_1) be the fundamental positive integral solution of Pell's equation $u^2 - kv^2 = 1$. If u_1 and v_1 are of opposite parity, we obtain infinitely many (but not necessarily all) positive integral solutions (x, y) by taking

$$2y + 1 + (2x+1)\sqrt{k} = (1+\sqrt{k})(u_1 + v_1\sqrt{k})^j, \quad j = 1, 2, 3, \dots$$

If u_1 and v_1 are both odd (which can occur only when $k \equiv 0 \pmod 8$), we set

$$u + v\sqrt{k} = (u_1 + v_1\sqrt{k})^2,$$

and since u_1 is odd and v_1 is even, we get infinitely many positive integral solutions x, y by taking

$$2y + 1 + (2x + 1)\sqrt{k} = (1 + \sqrt{k})(u + v\sqrt{k})^j, \quad j = 1, 2, 3, \ldots$$

This completes the proof of the theorem. □

The equation (7.2.2) is also studied in [34].

It is interesting to see what is the asymptotic density of "composite" triangular numbers among all triangular numbers. In [214] it is shown that this density is zero. More specifically, if $F(n)$ denotes the number of triples a, b, c such that

$$T_a T_b = T_c, \quad 1 < a \leq b < c \leq n \tag{7.2.3}$$

we will show that

$$F(n) < 4n^{3/4}. \tag{7.2.4}$$

Denote $g(x) = A\sqrt{x^2 - d^2}$ for $x \geq d$, where A and d are given positive numbers. Suppose h is a fixed positive number. Then

$$m_h(x) = \frac{1}{2x}[g(x + h) + g(x - h)]$$

is an increasing function of x for $x \geq d + h$.

Clearly,

$$m_h(x) = \frac{g(x + h)^2 - g(x - h)^2}{2x\{g(x + h) - g(x - h)\}} = \frac{2A^2 h}{g(x + h) - g(x - h)}.$$

Thus it suffices to show that $g(x + h) - g(x - h)$ is a decreasing function of x for $x \geq d + h$. But for $x > d + h$ we have

$$g'(x + h) - g'(x - h) = \frac{A(x + h)}{\sqrt{(x + h)^2 - d^2}} - \frac{A(x - h)}{\sqrt{(x - h)^2 - d^2}} < 0,$$

since the derivative of $x(x^2 - d^2)^{-1/2}$ is $-d^2(x^2 - d^2)^{-3/2}$.

Because $F(n) = 0$ for $n \leq 7$ we may assume $n \geq 8$. For a given a with $1 < a < n$, let $s(a, n)$ denote the number of pairs (b, c) satisfying (7.2.3). If $b \geq a > 2^{1/4}n^{1/2}$, clearly $T_b \geq T_a > \frac{1}{2} \cdot 2^{1/4}n^{1/2}(2^{1/4}n^{1/2} + 1)$ and, hence $T_a T_b > \frac{1}{2}(n + 2^{3/4})$.

Thus $s(a, n) = 0$ if $a > 2^{1/4}n^{1/2}$, and so

$$F(n) = \sum_{a=2}^{\lceil 2^{1/4}n^{1/2} \rceil} s(a, n).$$

Suppose that a and n are fixed and $s(a,n) > 0$. Set $K = T_a$. Then the equation $T_a T_b = t_c$ is equivalent to $K(b^2 + b) = c^2 + c$ or

$$K\{(2b+1)^2 - 1\} = (2c+1)^2 - 1.$$

Set $u = 2b + 1$, $v = 2c + 1$. Because $v^2 = (2c+1)^2 \leq (2n+1)^2$, we have

$$u^2 = \frac{v^2 - 1}{K} + 1 \leq \frac{4n^2 + 4n}{K} + 1 = 8\frac{n(n+1)}{a(a+1)} + 1 < 8\frac{n^2}{a^2} + 1 < 9\frac{n^2}{a^2},$$

so

$$u < 3n/a. \tag{7.2.5}$$

On the other hand, $u = 2b + 1 \geq 2a + 1 = \sqrt{8K+1} > \sqrt{2K}$ and

$$v^2 = Ku^2 - K + 1 > K\{(2b+1)^2 - 1\} \geq K\{(2a+1)^2 - 1\} = 8K^2,$$

hence

$$0 < \sqrt{K}u - v = \frac{K-1}{\sqrt{K}u + v} < \frac{K}{2\sqrt{2}K + 2\sqrt{2}K}$$

or

$$0 < \sqrt{K}u - v < 1/(4\sqrt{2}). \tag{7.2.6}$$

Now suppose (b_i, c_i), $i = 1, 2, \ldots, s$ are the solutions to $KT_b = T_c$ with $a \leq b_j < c_j \leq n$ and $b_1 < b_2 < \cdots < b_s$, where $s = s(a,n)$. Set $u_i = 2b_i + 1$ and $v_i = 2c_i + 1$ for $i = 1, 2, \ldots, s$. We claim that $u_{i+1} - u_i \neq u_{j+1} - u_j$ for $1 \leq i < j \leq s - 1$.

Suppose to the contrary that $u_{i+1} - u_i = u_{j+1} - u_j$ for some pair (i,j) with $1 \leq i < j \leq s - 1$. From (7.2.6) we have

$$-1/(4\sqrt{2}) < (\sqrt{K}u_{i+1} - v_{i+1}) - (\sqrt{K}u_i - v_i) < 1/(4\sqrt{2}),$$

so

$$\sqrt{K}(u_{i+1} - u_i) - (v_{i+1} - v_i) = \theta_i,$$

where $|\theta_i| < 1/(4\sqrt{2})$. Similarly,

$$\sqrt{K}(u_{j+1} - u_j) - (v_{j+1} - v_j) = \theta_j,$$

where $|\theta_j| < 1/(4\sqrt{2})$. Hence

$$v_{i+1} - v_i + \theta_i = \sqrt{K}(u_{i+1} - u_i) = \sqrt{K}(u_{j+1} - u_j) = v_{j+1} - v_j + \theta_j,$$

hence

$$[(v_{j+1} - v_j) - (v_{i+1} - v_i)] = |\theta_i - \theta_j| < 1/(2\sqrt{2}).$$

Because the left-hand side is an integer, we have

$$v_{i+1} - v_i = v_{j+1} - v_j. \qquad (7.2.7)$$

On the other hand $(u_i, v_i), (u_{i+1}, v_{i+1}), (u_j, v_j), (u_{j+1}, v_{j+1})$ are points with positive integral coordinates lying on the hyperbola $y^2 = Kx^2 - (K - 1)$ and satisfying the conditions $u_{i+1} - u_i = u_{j+1} - u_j > 0, u_i < u_j$. Further,

$$\frac{v_{i+1} - v_i}{v_{j+1} - v_j} = \frac{K(u_{i+1}^2 - u_i^2)/(v_{i+1} + v_i)}{K(u_{j+1}^2 - u_j^2)/(v_{j+1} + v_j)} = \frac{(v_{j+1} + v_j)/(u_{j+1} + u_j)}{(v_{i+1} + v_i)/(u_{i+1} + u_i)}. \qquad (7.2.8)$$

Applying the monotonicity of function m_h with $2h = u_{i+1} - u_i = u_{j+1} - u_j$ and $g(x) = \sqrt{Kx^2 - (K - 1)}$, we find that

$$(v_{j+1} + v_j)/(u_{j+1} + u_j) > (v_{i+1} + v_i)/(u_{i+1} + u_i),$$

and then (7.2.8) gives $v_{i+1} - v_j > v_{j+1} - v_j$. But this contradicts (7.2.7) and so our assumption that $u_{i+1} - u_i = u_{j+1} - u_j$ is untenable.

Thus we have shown that the gaps $u_2 - u_1, u_3 - u_2, \ldots, u_s - u_{s-1}$ are $s - 1$ different even positive integers. Hence,

$$u_s - u_1 = (u_2 - u_1) + (u_3 - u_2) + \cdots + (u_s - u_{s-1})$$
$$\geq 2 + 4 + \cdots + 2(s - 1) = s(s - 1).$$

Combining this with (7.2.5), we obtain

$$3n/a > u_s > u_s - u_1 \geq (s - 1)^2,$$

so

$$s(a, n) < 1 + \sqrt{3n/a}.$$

Hence

$$F(n) = \sum_{a=2}^{[2^{1/4}n^{1/2}]} s(a, n) < \sum_{a=2}^{[2^{1/4}n^{1/2}]} (1 + \sqrt{3n/a})$$

$$< 2^{1/4}n^{1/2} + \sqrt{3n} \int_1^{2^{1/4}n^{1/2}} t^{-1/2} dt$$

$$< \sqrt{3n} \int_0^{2^{1/4}n^{1/2}} t^{-1/2} dt < 4n^{3/4}.$$

Thus (7.2.4) is proved. □

The following result was proven in [170].

Theorem 7.2.3. *The equation*

$$T_m = T_n T_p \tag{7.2.9}$$

is solvable for infinitely many triples (m, n, p), $p \geq 2$, *and unsolvable for infinitely many triples* (m, n, p).

Proof. For the first part, we choose $p = 2$. The equation (7.2.9) becomes $T_m = 3T_n$. From Theorem 7.2.2 it follows that the last equation has infinitely many solutions.

For the second part, let m be an odd prime number. The equation (7.2.9) is equivalent to

$$2m(m + 1) = n(n + 1)p(p + 1).$$

Without loss of generality, we may assume that $m|n$ or $m|n + 1$, i.e., $n = km$ or $n + 1 = km$.

Since $p(p + 1) \geq 6$, in the first case we obtain

$$2(m + 1) = k(km + 1)p(p + 1) \geq 6(m + 1),$$

a contradiction. In the second case, when $n = km - 1$, we have

$$2(m + 1) = (km - 1)kp(p + 1) \geq 6(m - 1)$$

which is a contradiction, as well. It follows that equation (7.2.9) is not solvable. □

7.2.2 Rational Numbers Representable as $\dfrac{T_m}{T_n}$

The following results have been proven in [170]. The proof of the first result is based on some results contained in our papers [13] and [14].

Theorem 7.2.4. *If r is a positive rational number and $\sqrt{r}$ is irrational, then there exist positive integers m, n such that*

$$r = \frac{T_m}{T_n}. \tag{7.2.10}$$

Proof. Let $r = \dfrac{p}{q}$, where p, q are relatively prime positive integers. Then (7.2.10) is equivalent to $\dfrac{m(m + 1)}{n(n + 1)} = \dfrac{p}{q}$, i.e., $\dfrac{(2m + 1)^2 - 1}{(2n + 1)^2 - 1} = \dfrac{p}{q}$. Letting $2m + 1 = x$ and $2n + 1 = y$ yields

$$qx^2 - py^2 = q - p. \tag{7.2.11}$$

The irrationality of $\sqrt{r}$ implies that pq is not a perfect square.

Since (7.2.11) is solvable (it has solution $x = y = 1$), from Theorem 4.5.1 it follows that it has infinitely many solutions.

If $(u_k, v_k)_{k \geq 0}$ is the general solution to the Pell's equation $u^2 - pqv^2 = 1$, then $u_k + v_k\sqrt{pq} = (u_0 + v_0\sqrt{pq})^k$, $k \geq 1$. It follows that $u_1 = u_0^2 + pqv_0^2$, $v_1 = 2u_0v_0$.

From Theorem 4.5.1, $(X_k, Y_k)_{k \geq 0}$, where $X_k = u_k + pv_k$, $Y_k = u_k + qv_k$, are solutions to the equation (7.2.11). Since $u_1^2 - pqv_1^2 = 1$ and v_1 is even, it follows that u_1 is odd. Hence X_1 and Y_1 are both odd and we can choose $m = \dfrac{1}{2}(X_1 - 1)$ and $n = \dfrac{1}{2}(Y_1 - 1)$. □

Theorem 7.2.5. *Among the positive rational numbers r for which $\sqrt{r}$ is rational, infinitely many are representable in the form (7.2.10) and infinitely many are not.*

Proof. Let p be an odd integer and let $r = (2p)^2$. Choosing $m = p^2 - 1$ and $n = \dfrac{p-1}{2}$, we obtain $r = \dfrac{T_m}{T_n}$.

If p is an odd prime, we will prove that $r = \left(\dfrac{p+1}{p-1}\right)^2$ is not of the form (7.2.10).

Indeed $r = \dfrac{T_m}{T_n}$ would imply $\dfrac{m(m+1)}{n(n+1)} = \dfrac{(p+1)^2}{(p-1)^2}$. Setting $2m + 1 = x$ and $2n + 1 = y$, we have $\dfrac{x^2 - 1}{y^2 - 1} = \dfrac{(p+1)^2}{(p-1)^2}$, $x, y \geq 3$. The last equality is equivalent to

$$\left(\frac{p-1}{2}x - \frac{p+1}{2}y\right)\left(\frac{p-1}{2}x + \frac{p+1}{2}y\right) = -p$$

and so

$$\frac{p-1}{2}x - \frac{p+1}{2}y = -1 \text{ and } \frac{p-1}{2}x + \frac{p+1}{2}y = p,$$

which yields $x = 1$ and $y = 1$, a contradiction. □

7.2.3 When Is $\dfrac{T_m}{T_n}$ a Perfect Square?

In this subsection we are interested in finding all pairs (m, n) for which the ratio of triangular numbers T_m and T_n is the square of an integer.

In [140] it is shown that pairs $(4n(n + 1), n)$, $n \geq 1$, satisfy the above property. In the recent paper [101] all pairs (m, n) are determined by using a suitable Pell's equation.

The relation $\dfrac{T_m}{T_n} = q^2$ is equivalent to

$$\frac{(2m+1)^2 - 1}{(2n+1)^2 - 1} = q^2.$$

Using now the result and notation in Remark 5), subsection 5.6.2, we obtain $2m_k + 1 = x_k$, $q_k = z_k$, $k \geq 0$. It follows that $m_k = \dfrac{x_k - 1}{2}$, $q_k = z_k$, where

$$\begin{cases} x_k = \dfrac{1}{2}\left[\left(2n+1+2\sqrt{n(n+1)}\right)^k + \left(2n+1-2\sqrt{n(n+1)}\right)^k\right] \\[4mm] z_k = \dfrac{1}{4\sqrt{n(n+1)}}\left[\left(2n+1+2\sqrt{n(n+1)}\right)^k - \left(2n+1-2\sqrt{n(n+1)}\right)^k\right] \end{cases}$$

$k \geq 0$.

It is clear that x_k is odd for all k, hence all such pairs (m, n) are given by $(m_k, n)_{k \geq 0}$, where n is an arbitrary positive integer.

7.3   Polygonal Numbers

The kth polygonal number of order n (or the kth n-gonal number) P_k^n is given by the equation

$$P_k^n = \frac{k}{2}[(n-2)(k-1) + 2].$$

Diophantus (c. 250 A.D.) noted that if the arithmetic progression with first term 1 and common difference $n - 2$ is considered, then the sum of the first k terms is P_k^n. The usual geometric realization, from which the name derives, is obtained by considering regular polygons with n sides sharing a common angle and having points at equal distances along each side with the total number of points being P_k^n.

The first forty pages of Dickson's *History of Number Theory*, Vol. II, are devoted to results on polygonal numbers.

In [201] it is shown that there are infinitely many triangular numbers which at the same time can be written as the sum, the difference, and the product of other triangular numbers. It is easy to show that $4(m^2 + 1)^2$ is the sum, difference, and product of squares. Since then, several authors have proved similar results for sums and differences of other polygonal numbers. In [85] are considered pentagonal numbers, in [162] and [163] are considered hexagonal and septagonal numbers, and in [6] it is proved that for any n infinitely many n-gonal numbers can be written as the sum and difference of other n-gonal numbers. Although [85] gives several examples of pentagonal numbers written as the product of two other pentagonal numbers, the existence of an infinite class was left in doubt.

In this section we show that for every n there are infinitely many n-gonal numbers that can be written as the product of two other n-gonal numbers, and in fact show how to generate infinitely many such products. We suspect that our method does not generate all of the solutions for every n, but we have not tried to prove this. Moreover, except for $n = 3$ and 4, it is still not known whether there are infinitely many n-gonal numbers which at the same time can be written as the sum, difference, and product of n-gonal numbers.

Our proof uses the theory of the Pell equation. We also use a result on the existence of infinitely many solutions of a Pell equation satisfying a congruence condition, given that one solution exists satisfying the congruence condition. Next we note some facts about the Pell equation and prove this latter result. Then we prove the theorem on products of polygonal numbers.

In what follows, $\mathbb{Z}_+$ denotes the set of positive integers and $(a, b) \equiv (c, d)$ (mod m) means that $a \equiv c$ and $b \equiv d$ (mod m).

Theorem 7.3.1. *If $D \in \mathbb{Z}_+$ is not a square, then for any $m \in \mathbb{Z}_+$ there are infinitely many integral solutions to the Pell's equation $u^2 - Dv^2 = 1$, with $(u, v) \equiv (1, 0)$ (mod m).*

Proof. Suppose (u_1, v_1) is the fundamental solution to Pell's equation

$$u^2 - Dv^2 = 1$$

and that $(u_j, v_j)_{j \geq 1}$ is the general solution given by

$$u_j + v_j \sqrt{D} = (u_1 + v_1 \sqrt{D})^j, \quad j \geq 1.$$

Since there are only m^2 distinct ordered pairs of integers modulo m, there must be $j, l \in \mathbb{Z}_+$ such that $(u_j, v_j) \equiv (u_l, v_l)$ (mod m). We notice that, for any $k \geq 2$,

$$u_k + v_k \sqrt{D} = (u_1 + v_1 \sqrt{D})(u_{k-1} + v_{k-1} \sqrt{D})$$

so

$$u_k = u_1 u_{k-1} + D v_1 v_{k-1} \text{ and } v_k = v_1 u_{k-1} + u_1 v_{k-1}$$

(see also Section 3.2).

Applying these equations to the above congruence, we deduce

$$
\begin{aligned}
u_1 u_{j-1} + D v_1 v_{j-1} &\equiv u_1 u_{l-1} + D v_1 v_{l-1} \pmod{m} \text{ and} \\
v_1 u_{j-1} + u_1 v_{j-1} &\equiv v_1 u_{l-1} + u_1 v_{l-1} \pmod{m}.
\end{aligned}
\tag{7.3.1}
$$

From (7.3.1) it follows $(u_1^2 - Dv_1^2)v_{j-1} \equiv (u_1^2 - Dv_1^2)v_{l-1} \pmod{m}$. Since $u_1^2 - Dv_1^2 = 1$, we obtain $v_{j-1} \equiv v_{l-1} \pmod{m}$.

Similarly, from (7.3.1) we obtain $u_{j-1} \equiv v_{l-1} \pmod{m}$, so $(u_{j-1}, v_{j-1}) \equiv (u_{l-1}, v_{l-1}) \pmod{m}$.

We can conclude that for $i = |j - l|$, $(1, 0) = (u_0, v_0) \equiv (u_{si}, v_{si}) \pmod{m}$, for any $i \in \mathbb{Z}_+$.   □

Theorem 7.3.2. *If $a, b, m, D \in \mathbb{Z}_+$, D is not a square, and the general Pell's equation $u^2 - Dv^2 = M$ has a solution (u^*, v^*) with $(u^*, v^*) \equiv (a, b) \pmod{m}$, then it has infinitely many solutions $(u_k^*, v_k^*)_{k \geq 1}$ such that $(u_k^*, v_k^*) \equiv (a, b) \pmod{m}$.*

Proof. Let $(u_k, v_k)_{k \geq 1}$ be the solutions to the Pell's equation $u^2 - Dv^2 = 1$, guaranteed by Theorem 7.3.1, i.e., $(u_k, v_k) \equiv (1, 0) \pmod{m}$. Then the solution $(u_k^*, v_k^*)_{k \geq 1}$ to the general Pell's equation obtained from these solutions are such that

$$u_k^* = u^* u_k + Dv^* v_k \equiv a \cdot 1 + Db \cdot 0 \equiv a \pmod{m}$$

and

$$v_k^* = v^* u_k + u^* v_k \equiv b \cdot 1 + a \cdot 0 \equiv b \pmod{m}, \quad k \geq 1$$

(see also Section 4.1).   □

The following Corollary follows by taking m in Theorem 7.3.2 to be the least common multiple of m_1 and m_2.

Corollary 7.3.3. *If $a, b, m_1, m_2, D \in \mathbb{Z}_+$, D is not a square, and $a^2 - Db^2 = M$, then there are infinitely many solutions to the general Pell's equation $u^2 - Dv^2 = M$ with $u \equiv a \pmod{m_1}$ and $v \equiv b \pmod{m_2}$.*

Next we show that any nonsquare n-gonal number is infinitely often the quotient of two n-gonal numbers (see [68]). The theorem that n-gonal products are infinitely often n-gonal and a remark on the solvability of a related equation follow.

Theorem 7.3.4. *If the n-gonal number $P = P_s$ is not a square, then there exist infinitely many distinct pairs (P_x, P_y) of n-gonal numbers such that*

$$P_x = P_s P_y. \tag{7.3.2}$$

Proof. Recalling that $P_x = \dfrac{x}{2}[(n - 2)(x - 1) + 1]$ and setting $n - 2 = r$, equation (7.3.2) becomes

$$rx^2 - (r - 2)x = P[ry^2 - (r - 2)y].$$

Multiplying by $4r$ to complete the square gives

$$(2rx - (r - 2))^2 - (r - 2)^2 = P[(2ry - (r - 2))^2 - (r - 2)^2].$$

Setting

$$u = 2rx - (r - 2), \quad v = 2ry - (r - 2) \qquad (7.3.3)$$

we get the general Pell's equation

$$u^2 - Pv^2 = M, \qquad (7.3.4)$$

with $M = (r - 2)^2 - P(r - 2)^2$.

Thus, in order to ensure infinitely many solution (x, y) to (7.3.2), it suffices to have infinitely many solutions (u, v) to (7.3.4) for which the pair (x, y) obtained from (7.3.3) is integral. Put another way, it suffices to show the existence of infinitely many solutions (u^*, v^*) to (7.3.4) for which the congruence

$$(u^*, v^*) \equiv (-(r - 2), -(r - 2)) \equiv (r + 2, r + 2) \pmod{2r}$$

holds.

But notice that, since $P_1 = 1$, a particular solution of (7.3.2) is $x = s$, $y = 1$, and these values of x and y give $u = (2s - 1)r + 2$, $v = r + 2$, as a particular solution of (7.3.4). Thus, we have a solution (u^*, v^*) of (7.3.4) with $(u^*, v^*) \equiv (r + 2, r + 2)$ $\pmod{2r}$. Theorem 7.3.2 guarantees the infinitely many solutions we are seeking. $\square$

Our final result is now a straightforward corollary.

Theorem 7.3.5. *For any $n \geq 3$, there are infinitely many n-gonal numbers which can be written as a product of two other n-gonal numbers.*

Proof. The case $n = 4$ is trivial. By Theorem 7.3.4, we need only show that P_s is not a square for some s. But for $n \neq 4$, at least one of $P_2 = n$ and $P_9 = 9(4n - 7)$ is not a square. $\square$

Remarks. 1) Trying to prove that

$$P_k = \frac{k}{2}[(n - 2)(k - 1) + 2] = P_x P_y$$

infinitely often by setting $P_x = k$ and

$$P_y = \frac{1}{2}[(n - 2)(P_x - 1) + 2]$$

and solving the corresponding Pell's equation that results works if $n \neq 2t^2 + 2$, and thus, for these values of n, there are infinitely many solutions to the equation $P_p = P_x P_y$.

2) There are 51 solutions of $P_x^3 = P_s^3 P_y^3$ with $P_x^3 < 10^6$. There are 43 solutions of $P_x^n = P_s^n P_y^n$ with $5 \leq n \leq 36$ and $P_x^n < 10^6$. For $36 < n \leq 720$, there are no solutions with $P_x^n < 10^6$.

3) In [107] are considered the simultaneous equations $P_x^n = P_y^m = P_z^q$, where m, n, q, x, y, z are positive integers. By reducing these to systems of simultaneous Pell equations, one can show that if (m, n, q) is not a permutation of $(3, 6, k)$ (for $k > 3$), then all solutions of the above system of equations have $\max\{x, y, z\} < c$, where c is an effectively computable constant depending only on m, n and q. In fact, the remaining case may also be easily analyzed, upon noting the reduction to

$$Z^2 - jX^2 = (j - 1)(j - 4),$$

if we take $(m, n, q) = (3, 6, j + 2)$. If j is a square, this equation has at most finitely many solutions, while, if $j > 1$ is not a square, it has infinitely many, corresponding to classes of the given Pell's equation, upon noting that $(Z, X) = (j - 2, 1)$ gives one such solution.

7.4   Powerful Numbers

Define a positive integer r to be a *powerful number* if p^2 divides r whenever the prime p divides r. The following list contains all powerful numbers between 1 and 1000: 1, 4, 8, 9, 16, 25, 27, 32, 36, 49, 64, 72, 81, 100, 108, 121, 125, 128, 144, 169, 196, 200, 216, 225, 243, 256, 288, 289, 324, 343, 361, 392, 400, 432, 441, 484, 500, 512, 529, 576, 625, 648, 675, 676, 729, 784, 800, 841, 864, 900, 961, 968, 972, 1000. Let $k(x)$ denote the number of powerful numbers not exceeding x. Following [77] we show that

$$\lim_{x \to \infty} \frac{k(x)}{\sqrt{x}} = c,$$

with the constant $c = \dfrac{\zeta(3/2)}{\zeta(3)}$, where ζ is the well-known Riemann zeta function.

We also prove that there are infinitely many pairs of consecutive powerful integers, such as 8, 9 and 288, 289. We conclude with some results and conjectures concerning the gaps between powerful numbers.

7.4.1 The Density of Powerful Numbers

Let

$$F(s) = \prod_p (1 + p^{-2s} + p^{-3s} + \ldots) = \prod_p \left(1 + \frac{1}{p^s(p^s - 1)}\right) \qquad (7.4.1)$$

where the products are extended over all primes p. It is evident that

$$F(s) = \sum_{r \in K} r^{-s}, \qquad (7.4.2)$$

where K is the set of powerful numbers. Then, the sum of the reciprocals of the powerful numbers,

$$F(1) = \sum_{r \in K} \frac{1}{r} = \prod_p \left(1 + \frac{1}{p(p - 1)}\right), \qquad (7.4.3)$$

is seen to be convergent (see [136–138] for the theory of convergent series).

To estimate $k(x)$, the number of powerful numbers up to x, we observe first that $k(x) \geq [\sqrt{x}]$, since every perfect square is powerful. Next, we observe that every powerful number r can be represented as a perfect square n^2 (including the case $n = 1$) times a perfect cube m^3 (including $m = 1$), and that this representation is unique if we require m to be square-free. That is, we set m equal to the product of those primes having odd exponents in the canonical factorization of r into powers of distinct primes, and the representation $r = n^2 m^3$ is then unique.

Thus,

$$k(x) = \#(n^2 m^2 \leq x, \mu(m) \neq 0) = \sum_{m=1}^{\infty} \mu^2(m) \left[\left(\frac{x}{m^3}\right)^{1/2}\right] \sim cx^{1/2}, \ x \to \infty, \qquad (7.4.4)$$

where

$$\sum_{m=1}^{\infty} \mu^2(m) m^{-3/2} < \zeta(3/2) < \infty. \qquad (7.4.5)$$

Explicitly,

$$c = \prod_p (1 + p^{-3/2}) = \prod_p (1 - p^{-3})/(1 - p^{-3/2}) = \zeta(3/2)\zeta(3), \qquad (7.4.6)$$

where $\zeta(s)$ is the Riemann zeta function (see [213]). This evaluation of c comes from setting $s = 3/2$ in the identity

$$\sum_{m=1} \frac{\mu^2(m)}{m^s} = \prod_p \left(1 + \frac{1}{p^s}\right)$$

$$= \prod_p \frac{1 - p^{-2s}}{1 - p^{-s}}$$

$$= \prod_p (1 - p^{-2s}) \Big/ \prod_p (1 - p^{-s})$$

$$= \zeta(s)/\zeta(2s) \qquad (7.4.7)$$

for all $Re(s) > 1$, where $\zeta(s) = \sum_{n=1}^{\infty} n^{-s} = \prod_p (1 - p^{-s})^{-1}$ for $Re(s) > 1$.

For purposes of estimation, we have the inequalities

$$cx^{1/2} \geq k(x) \geq cx^{1/2} - 3x^{1/3} \text{ for } x \geq 1, \qquad (7.4.8)$$

because $cx^{1/2} = \sum_{m=1}^{\infty} \mu^2(m)(x/m^3)^{1/2} \geq \sum_{m=1}^{\infty} \mu^2(m)[(x/m^3)^{1/2}] = k(x)$, and

$$cx^{1/2} - k(x) = \sum_{m=1}^{\infty} |\mu(m)| \left\{ \left(\frac{x}{m^3}\right)^{1/2} - \left[\left(\frac{x}{m^3}\right)^{1/2}\right] \right\}$$

$$\leq \sum_{m=1}^{[x^{1/3}]-1} |\mu(m)| \cdot 1 + \sum_{m=[x^{1/3}]}^{\infty} |\mu(m)| \left(\frac{x}{m^3}\right)^{1/2}$$

$$\leq ([x^{1/3}] - 1) + \left(1 + \sqrt{x} \int_{[x^{1/3}]}^{\infty} u^{-3/2} du\right)$$

$$\leq x^{1/3} + 2x^{1/2}x^{-1/6} = 3x^{1/3}.$$

Numerically, $c = 2 \cdot 173\ldots$

We have the further identities:

$$F(s) = \sum_{r \in K} (1/r^s)$$

$$= \sum_{n=1}^{\infty} n^{-2s} \sum_{m=1}^{\infty} \mu^2(m)m^{-2s}$$

$$= \sum_{i=1}^{\infty} t^{-2s} \sum_{m|t} |\mu(m)|/m^s$$

$$= \sum_{t=1}^{\infty} t^{-2s} \prod_{p|t} (1 + p^{-s}), \tag{7.4.9}$$

where we used the substitution $t = mn$;

$$F(s) = \sum_{n=1}^{\infty} n^{-2s} \sum_{m=1}^{\infty} \mu^2(m) m^{-2s} = \zeta(2s)\zeta(3s)/\zeta(6s), \tag{7.4.10}$$

and

$$F(1) = \zeta(2)\zeta(3)/\zeta(6) = \sum_{i=1}^{\infty} \Psi(t)/t^3, \tag{7.4.11}$$

where

$$\Psi(t) = t \prod_{p|t} \left(1 + \frac{1}{p}\right), \tag{7.4.12}$$

by setting $s = 1$ in the previous identities (7.4.9) and (7.4.10).
   Since $\zeta(2) = \pi^2/6$ and $\zeta(6) = \pi^6/945$, we observe

$$F(1) = \frac{315}{2\pi^4} \zeta(3). \tag{7.4.13}$$

7.4.2 Consecutive Powerful Numbers

Four consecutive integers cannot all be powerful, since one of them is twice an odd number. No example of three consecutive powerful numbers is known, unless one is willing to accept -1, 0, 1. If such an example exists, it must be of the form

$$4k - 1, \quad 4k, \quad 4k + 1.$$

No case of $4k - 1$ and $4k + 1$ both being powerful is known. In fact, the only known example of consecutive odd numbers $2k - 1$ and $2k + 1$ both being powerful is $2k - 1 = 25$, $2k + 1 = 27$.
   There are two infinite families of examples where two consecutive integers are powerful which correspond to the solutions of the Pell equations $x^2 - 2y^2 = 1$ and $x^2 - 2y^2 = -1$.
   Let x_1, y_1 satisfy $x_1^2 - 2y_1^2 = \pm 1$. Then $8x_1^2 y_1^2 = A$ and $(x_1^2 + 2y_1^2)^2 = B$ are consecutive powerful numbers. The following table gives several examples of consecutive powerful numbers from solutions of the equations $x^2 - 2y^2 = \pm 1$.

x	y	A	B
1	1	$8 = 2^3$	$9 = 3^2$
3	2	$288 = 2^5 \cdot 3^2$	$289 = 17^2$
7	5	$9800 = 2^3 \cdot 5^2 \cdot 7^2$	$9801 = 3^4 \cdot 11^2$
17	12	$332928 = 2^5 \cdot 3^2 \cdot 17^2$	$332929 = 577^2$
$\sqrt{B_0}$	$\sqrt{A_0/2}$	$4A_0B_0$	$4A_0B_0 + 1$

If A and $B = A + 1$ are consecutive powerful numbers, and if B is a perfect square, $B = u^2$, then $A = (u - 1)(u + 1)$. If u is even, then $(u - 1, u + 1) = 1$, and both $u - 1$ and $u + 1$ are odd powerful numbers. As already remarked, the only known instance of this occurrence is $u - 1 = 25, u + 1 = 27$, leading to the isolated example $A = 675 = 3^3 \cdot 5^2, B = 676 = 2^2 \cdot 13^2$. If u is odd, then $(u - 1)/2$ and $(u + 1)/2$ are consecutive integers, with $((u - 1)/2, (u + 1)/2) = 1$. For $u^2 - 1$ to be powerful, $(u - 1)/2$ and $(u + 1)/2$ must be a powerful odd number and twice a powerful number, in either order. The two Pell equations produce examples in both orders. However, an example satisfying neither of these Pell equations is also known, with $(u - 1)/2 = 242 = 2 \cdot 11^2$ and $(u + 1)/2 = 243 = 3^5$. This leads to $A = 235.224 = 2^3 \cdot 3^5 \cdot 11^2$ and $B = 235.225 = 5^2 \cdot 97^2$.

Whenever A and B are consecutive powerful numbers, so too are $A' = 4AB$ and $B' = 4AB + 1 = (2A + 1)^2$. The solution $x_0 = 1, y_0 = 1$, of $x^2 - 2y^2 = -1$ generates all solutions of the Pell equations $x^2 - 2y^2 = \pm 1$, in the sense that $x_n + y_n\sqrt{2} = (x_0 + y_0\sqrt{2})^n$ yields the complete set of solutions (x_n, y_n) such that $x_n^2 - 2y_n^2 = \pm 1$.

Note that the consecutive powerful numbers $A = 675, B = 676$, come from the solution $x = 26, y = 15$, of the Pell equation $x^2 - 3y^2 = 1$, with $A = 3y^2$ and $B = x^2$. Similarly, the example $A = 235.224, B = 235.225$ of consecutive powerful numbers arises from the Pell equation $x^2 - 6y^2 = 1$ with $x = 485, y = 198$. More generally, any solution (x_1, y_1) of the Pell equation $x^2 - dy^2 = \pm 1$, with the extra condition that $d | y_1^2$, leads to an infinite family of consecutive powerful numbers, starting with $A_1 = x_1^2, B_1 = dy_1^2 = A_1 \pm 1$, and continuing with $A_n = x_n^2, B_n = dy_n^2$, where (x_n, y_n) are obtained from the computation $(x_1 + \sqrt{d}y_1)^n = x_n + \sqrt{d}y_n$.

Conversely, whenever we have two consecutive powerful integers, if one of them is a perfect square x^2, we can write the other in the form $n^2m^3 = my^2$, with m square-free, and we have a solution to the Pell equation $x^2 - my^2 = \pm 1$.

In all cases given thus far consecutive powerful numbers, the larger number is a perfect square. However, the Pell equation $x^2 - 5y^2 = -1$ with $5 | y$ leads to infinitely many powerful numbers $x^2 + 1 = 5y^2$, such as $x^2 = (682)^2 = 465124; 5y^2 = 5(305)^2 = 5^3 \cdot 61^2 = 465125$.

One example consisting of two consecutive powerful numbers where neither is a perfect square is given by $A = 23^3 = 12167$ and $B = 2^3 \cdot 3^2 \cdot 13^2 = 12168$. An interesting method based on the equation $ax^2 - by^2 = 1$ to generate such consecutive powerful numbers is presented in the paper [220]. For instance, in this paper is found $A = 7(2637362)^2 = 48689748233308$ and $B = 3(4028637)^2 = 48689748233307$.

7.4.3 Gaps Between Powerful Numbers

The set K of powerful numbers is closed under multiplication. Since there are infinitely many pairs of powerful numbers which differ by 1, there are infinitely many pairs of powerful numbers differing by r, for any $r \in K$.

Every positive integer not of the form $2(2b + 1)$ is difference of two powerful numbers in at least one way (specifically, as a difference of two perfect squares). For numbers of the form $2(2b + 1)$, $b \geq 0$, such representations may also exist. Thus:

$$2 = 3^3 - 5^2 \qquad\qquad 30 = 83^2 - 19^3$$
$$6 = ? \qquad\qquad 34 = ?$$
$$10 = 13^3 - 3^7 \qquad\qquad 38 = 37^2 - 11^3$$
$$14 = ? \qquad\qquad 42 = ?$$
$$18 = 19^2 - 7^3 = 3^2(3^3 - 5^2) \quad 46 = 17^2 - 3^5$$
$$22 = 7^2 - 3^3 = 47^2 - 3^7 \qquad 50 = 5^2(3^3 - 5^2)$$
$$26 = 3^3 - 1^2 = 7^2 \cdot 3^5 - 109^2 \quad 54 = 3^4 - 3^3 = 3^3(3^3 - 5^2) = 7^3 - 17^2.$$

If u and v are both powerful numbers, $(u, v) = 1$, and $a = u - v$, we say that a has a proper representation as a difference of powerful numbers. We observe that

$$2b + 1 = (b + 1)^2 - b^2$$
$$8c = (2c + 1)^2 - (2c - 1)^2$$

so that all odd numbers, as well as all multiples of 8, have proper representations. Among the numbers $2(2b + 1)$, $b = 0, 1, \ldots, 13$ for which representations were found, there were proper representations included in every case except $2(2b + 1) = 50$. Finally, for numbers $4(2b + 1)$, $b = 0, 1, \ldots, 12$, we observe the following proper representations:

$$4 = 5^3 - 11^2 \qquad\qquad 60 = ?$$
$$12 = 47^2 - 13^3 \qquad\qquad 68 = 3^3 \cdot 5^4 - 7^5$$
$$20 = ? \qquad\qquad 76 = 5^3 - 7^2$$
$$28 = ? \qquad\qquad 84 = ?$$
$$36 = ? \qquad\qquad 92 = ?$$
$$44 = 5^3 - 3^4 = 13^2 - 5^3 \quad 100 = 7^3 - 3^5$$
$$52 = ?$$

It is interesting that if u and $v = u + 4$ are both powerful, then so too are $u' = uv$ and $v' = u' + 4 = (u + 2)^2$. Thus, from the example $4 = 5^3 - 11^2$, an infinite number of proper representations of 4 are obtainable. It would be interesting to determine whether or not any numbers other than 1 and 4 have infinitely many proper representations.

Among the powerful numbers which are not perfect squares, the smallest difference known to occur infinitely often is 4. Specifically, the equation $3x^2 - 2y^2 = 1$ has infinitely many solutions for which $3|x$, such as $x = 9$, $y = 11$. For any such solution, we have $12x^2 - 8y^2 = 4$, where $12x^2$ and $8y^2$ are both powerful, and neither is a square. The only known instances where the difference between nonsquare powerful numbers is less than 4 are: $2^7 - 5^3 = 3$, and (as previously mentioned) $2^3 \cdot 3^2 \cdot 13^2 - 23^3 = 1$.

It has been conjectured that 6 cannot be represented in any way as a difference between two powerful numbers. It is further conjectured that there are infinitely many numbers which cannot be so represented. Other interesting properties and open problems concerning powerful numbers are mentioned in [141].

7.5   The Diophantine Face of a Problem Involving Matrices in $M_2(\mathbb{Z})$

Let R be a ring with identity. An element $a \in R$ is called *unit-regular* if $a = bub$ with $b \in R$ and a unit u in R, *clean* if $a = e+u$ with an idempotent e and a unit u, and *nil-clean* if $a = e + n$ with an idempotent e and a nilpotent n. A ring is *unit-regular* (or *clean*, or *nil-clean*) if all its elements are so. In [48], it was proved that *every unit-regular ring is clean*. However, in [103], it was noticed that this implication, *for elements*, fails. In the paper, plenty of unit-regular elements which are not clean are found among 2×2 matrices of the type $\begin{bmatrix} a & b \\ 0 & 0 \end{bmatrix}$ with integer entries.

While it is easy to prove that *any nil-clean ring is also a clean ring*, the question whether *nil-clean elements are clean*, was left open (see [63] and restated in [64]) for some 7 years. In this section, following the paper [29], we answer in the negative this question.

7.5.1   Nil-Clean Matrices in $M_2(\mathbb{Z})$

As this was done (in a special case) in [103], we investigate elements in the 2×2 matrix ring $M_2(\mathbb{Z})$. Since $\mathbb{Z}$ and direct sums of $\mathbb{Z}$ are not clean (not even exchange rings), it makes sense to look for elements which are not clean in this matrix ring.

We first recall some elementary facts.

Let R be an integral domain and $A \in M_n(R)$. Then A is a zero divisor if and only if $\det A = 0$. Therefore idempotents (excepting the identity matrix) and nilpotents have zero determinant.

For $A \in M_n(R)$, $rk(A) < n$ if and only if $\det A$ is a zero divisor in R. A matrix A is a unit in $M_n(R)$ if and only if $\det A \in U(R)$. Thus, *the units in $M_2(\mathbb{Z})$ are the 2×2 matrices of* $\det = \pm 1$.

Lemma 7.5.1. *Nontrivial idempotents in $M_2(\mathbb{Z})$ are matrices*

$$\begin{bmatrix} \alpha + 1 & u \\ v & -\alpha \end{bmatrix}$$

with $\alpha^2 + \alpha + uv = 0$.

Proof. One way follows by calculation. Conversely, notice that excepting I_2, such matrices are singular. Any nontrivial idempotent matrix in $M_2(\mathbb{Z})$ has rank 1. By Cayley–Hamilton Theorem, $E^2 - \text{tr}(E)E + \det(E)I_2 = 0$. Since $\det(E) = 0$ and $E^2 = E$ we obtain $(1 - \text{tr}(E)).E = 0_2$ and so, since there are no zero divisors in $\mathbb{Z}$, $\text{tr}(E) = 1$. □

Lemma 7.5.2. *Nilpotents in $M_2(\mathbb{Z})$ are matrices*

$$\begin{bmatrix} \beta & x \\ y & -\beta \end{bmatrix}$$

with $\beta^2 + xy = 0$.

Proof. One way follows by calculation. Conversely, just notice that nilpotent matrices in $M_2(\mathbb{Z})$ have the characteristic polynomial t^2 and so have trace and determinant equal to zero. □

Therefore *the set of all the nil-clean matrices* in $M_2(\mathbb{Z})$, which use a nontrivial idempotent in their nil-clean decomposition, is

$$\left\{ \begin{bmatrix} \alpha + \beta + 1 & u + x \\ v + y & -\alpha - \beta \end{bmatrix} \mid \alpha, \beta, u, v, x, y \in \mathbb{Z}, \alpha^2 + \alpha + uv = 0 = \beta^2 + xy \right\}.$$

Remarks. 1) Nil-clean matrices in $M_2(\mathbb{Z})$ which use a nontrivial idempotent, have the trace equal to 1. Otherwise, this is 2 or 0.
2) Since only the absence of nonzero zero divisors is (essentially) used, the above characterizations hold in any integral domain.

It is easy to discard the triangular case.

Proposition 7.5.3. *Upper triangular nil-clean matrices, which are neither unipotent nor nilpotent, are idempotent, and so (strongly) clean.*

Proof. Such upper triangular idempotents are $\begin{bmatrix} \alpha + 1 & u \\ 0 & -\alpha \end{bmatrix}$ with

$$-\det = \alpha^2 + \alpha = 0,$$

so have $\alpha \in \{-1, 0\}$, that is, $\begin{bmatrix} 1 & u \\ 0 & 0 \end{bmatrix}$ or $\begin{bmatrix} 0 & u \\ 0 & 1 \end{bmatrix}$. Upper triangular nilpotents have

the form $\begin{bmatrix} 0 & x \\ 0 & 0 \end{bmatrix}$, and so upper triangular nil-clean matrices have the form $\begin{bmatrix} 1 & u \\ 0 & 0 \end{bmatrix}$ or

$\begin{bmatrix} 0 & u \\ 0 & 1 \end{bmatrix}$. As noticed before, these are idempotent.   □

In the sequel we shall use the quadratic equation (3.1.1)

$$ax^2 + bxy + cy^2 + dx + ey + f = 0,$$

where a, b, c, d, e, f and are integers.

Denote $D =: b^2 - 4ac$, $g =: \gcd(b^2 - 4ac; 2ae - bd)$ and

$$\Delta =: 4acf + bde - ae^2 - cd^2 - fb^2.$$

Then the equation reduces to

$$-\frac{D}{g}Y^2 + gX^2 + 4a\frac{\Delta}{g} = 0$$

which (if $D > 0$) is a general Pell equation. Here

$$Y = 2ax + by + d \text{ and } X = \frac{D}{g}y + \frac{2ae - bd}{g}.$$

Notice that this equation may be also written as $-DY^2 + X^2 + 4a\Delta = 0$ replacing X by gX (and so $X = Dy + 2ae - bd$).

7.5.2   The General Case

In order to find a nil-clean matrix in $\mathcal{M}_2(\mathbb{Z})$ which is not clean, we need integers $\alpha, \beta, u, v, x, y$ with $\alpha^2 + \alpha + uv = 0 = \beta^2 + xy$, such that for every $\gamma, s, t \in \mathbb{Z}$, with $\gamma^2 + \gamma + st = 0$, the determinant

$$\det\left[\begin{bmatrix} \alpha + \beta - \gamma & u + x - s \\ v + y - t & -\alpha - \beta + \gamma \end{bmatrix}\right] = -(\alpha + \beta - \gamma)^2 - (u + x - s)(v + y - t) \notin \{\pm 1\}.$$

That is, subtracting any idempotent $\begin{bmatrix} \gamma + 1 & s \\ t & -\gamma \end{bmatrix}$ from

$$\begin{bmatrix} \alpha+1 & u \\ v & -\alpha \end{bmatrix} + \begin{bmatrix} \beta & x \\ y & -\beta \end{bmatrix},$$

the result should not be a unit in $M_2(\mathbb{Z})$.

Remark. Notice that above we have excepted the trivial idempotents. However, this will not harm since, in finding a counterexample, we ask for the nil-clean example not to be idempotent, nilpotent nor unit (and so not unipotent).

In the sequel, to simplify the writing, the following notations will be used: firstly, $m := 2\alpha + 2\beta + 1$ (m is odd and so nonzero) and $n := (u+x)(v+y) + (\alpha+\beta)^2 + 1$, and secondly, $r := \alpha + \beta$ and $\delta := r^2 + r + (v+y)(u+x)$. Then

$$m = 2r+1, \ n = (u+x)(v+y) + r^2 + 1 = \delta - r + 1.$$

This way an arbitrary nil-clean matrix which uses no trivial idempotents is now written

$$C = \begin{bmatrix} r+1 & u+x \\ v+y & -r \end{bmatrix}$$

and $\delta = -\det C$. To simplify the wording such nil-clean matrices will be called *nontrivial nil-clean.*

Theorem 7.5.4. *Let*

$$C = \begin{bmatrix} r+1 & u+x \\ v+y & -r \end{bmatrix}$$

be a nontrivial nil-clean matrix and let

$$E = \begin{bmatrix} \gamma+1 & s \\ t & -\gamma \end{bmatrix}$$

be a nontrivial idempotent matrix. With above notations, $C - E$ is invertible in $M_2(\mathbb{Z})$ with $\det(C - E) = 1$ if and only if

$$X^2 - (1+4\delta)Y^2 = 4(v+y)^2(2r+1)^2(\delta^2 + 2\delta + 2)$$

with

$$X = (2r+1)[-(1+4\delta)t + (2\delta+3)(v+y)]$$

and

$$Y = 2(v+y)^2 s + (2r^2 + 2r + 1 + 2\delta)t - (2\delta+3)(v+y).$$

Further, $C - E$ is invertible in $\mathcal{M}_2(\mathbb{Z})$ with $\det(C - E) = -1$ if and only if

$$X^2 - (1 + 4\delta)Y^2 = 4(v + y)^2(2r + 1)^2\delta(\delta - 2)$$

with

$$X = (2r + 1)[-(1 + 4\delta)t + (2\delta - 1)(v + y)]$$

and

$$Y = 2(v + y)^2 s + (2r^2 + 2r + 1 + 2\delta)t - (2\delta - 1)(v + y).$$

Proof. For given $\alpha, \beta, u, v, x, y$, $\det(C - E) = \pm 1$ amounts to a general inhomogeneous equation of the second degree with two unknowns, which we reduce to a canonical form, as mentioned in the previous section. Here are the details.

$$-\gamma^2 - st - (\alpha + \beta)^2 + 2(\alpha + \beta)\gamma + (v + y)s + (u + x)t - (u + x)(v + y)$$
$$= (2\alpha + 2\beta + 1)\gamma + (v + y)s + (u + x)t - (u + x)(v + y) - (\alpha + \beta)^2 = \pm 1.$$

The case $\det = 1$. Since

$$-m\gamma = (v + y)s + (u + x)t - (u + x)(v + y) - (\alpha + \beta)^2 - 1 = (v + y)s + (u + x)t - n,$$

we obtain from $(-m\gamma)^2 - m(-m\gamma) + m^2 st = 0$, the equation

$$[(v + y)s + (u + x)t - n]^2 - m[(v + y)s + (u + x)t - n] + m^2 st = 0,$$

or

$$(v + y)^2 \mathbf{s}^2 + [2(v + y)(u + x) + m^2]\mathbf{st} + (u + x)^2\mathbf{t}^2$$
$$-(m + 2n)(v + y)\mathbf{s} - (m + 2n)(u + x)\mathbf{t} + (m + n)n = 0.$$

Thus, with the notations of the previous section

$$\mathbf{a} = (v + y)^2, \; \mathbf{b} = [2(v + y)(u + x) + m^2], \; \mathbf{c} = (u + x)^2$$

and

$$\mathbf{d} = -(m + 2n)(v + y), \; \mathbf{e} = -(m + 2n)(u + x), \; \mathbf{f} = (m + n)n.$$

Further

$$\mathbf{D} = [2(v + y)(u + x) + m^2]^2 - 4(v + y)^2(u + x)^2$$

$$= m^4 + 4m^2(v+y)(u+x) = m^2[m^2 + 4(v+y)(u+x)],$$

$$2ae - bd = m^2(m+2n)(v+y) \text{ for } g = \gcd(D, 2ae - bd)$$

(notice that $m^2 | g$) and

$$\Delta = 4acf + bde - ae^2 - cd^2 - fb^2$$

$$= 4(v+y)^2(u+x)^2(m+n)n + [2(v+y)(u+x) + m^2](m+2n)^2(v+y)(u+x)$$

$$-(v+y)^2(m+2n)^2(u+x)^2 - (u+x)^2(m+2n)^2(v+y)^2$$

$$-(m+n)n[2(v+y)(u+x) + m^2]^2$$

$$= m^4[(v+y)(u+x) - (m+n)n].$$

The case det $= -1$. Formally exactly the same calculation, but n is slightly modified: here

$$n' = (u+x)(v+y) + (\alpha+\beta)^2 - 1,$$

i.e., $n' := n - 2$.

These equations reduce to the canonical form

$$gX^2 - \frac{D}{g}Y^2 = -4a\frac{\Delta}{g}$$

with

$$D = m^2[m^2 + 4(v+y)(u+x)],$$

$$g = \gcd(D, m^2(m+2n)(v+y)), \quad a = (v+y)^2$$

and

$$\Delta = m^4[(v+y)(u+x) - (m+n)n].$$

Since clearly $g = m^2 g'$, in the above equation we can replace D and Δ by $\dfrac{D}{m^2}$ and $\dfrac{\Delta}{m^2}$ (and $g = \gcd(m^2 + 4(v+y)(u+x); (m+2n)(v+y)))$, that is $D = m^2 + 4(v+y)(u+x)$ and $\Delta = m^2[(v+y)(u+x) - (m+n)n]$.

Further, this amounts to $g^2 X^2 - DY^2 = -4a\Delta$ and so we can eliminate g (by taking a new unknown: $X' = gX$). Hence we reduce to the equation

$$X^2 - [m^2 + 4(v+y)(u+x)]Y^2 = -4(v+y)^2 m^2[(v+y)(u+x) - (m+n)n].$$

which we can rewrite as

$$X^2 - (1 + 4\delta)Y^2 = 4(v + y)^2(2r + 1)^2(\delta^2 + 2\delta + 2).$$

Further, for det $= -1$, we obtain a similar equation replacing n by $n - 2$, i.e., $n = \delta - r - 1$:

$$X^2 - (1 + 4\delta)Y^2 = 4(v + y)^2(2r + 1)^2\delta(\delta - 2).$$

The linear systems in s and t corresponding to det $= 1$ and det $= -1$, are respectively:

$$\begin{cases} 2(v + y)^2 s + (2r^2 + 2r + 1 + 2\delta)t - (2\delta + 3)(v + y) = Y \\ \quad (2r + 1)[-(1 + 4\delta)t + (2\delta + 3)(v + y)] \qquad = X \end{cases}$$

for det $= 1$
(here $-(2r + 1)\gamma = (v + y)s + (u + x)t - n = (v + y)s + (u + x)t - \delta + r - 1$), and

$$\begin{cases} 2(v + y)^2 s + (2r^2 + 2r + 1 + 2\delta)t - (2\delta - 1)(v + y) = Y \\ \quad (2r + 1)[-(1 + 4\delta)t + (2\delta - 1)(v + y)] \qquad = X \end{cases}$$

for det $= -1$
(here $-(2r + 1)\gamma = (v + y)s + (u + x)t - n' = (v + y)s + (u + x)t - \delta + r + 1$). $\square$

7.5.3   The Example

Since $1 + 4\delta \geq 1$ if $\delta \geq 0$, in this case, from the general theory of Pell equations, it is known that the equations emphasized in Theorem 7.5.4 have infinitely many solutions, and so we cannot decide whether all the linear systems corresponding to these equations have (or not) integer solutions. However, if $\delta \leq -1$, then $1 + 4\delta < 0$ and we have elliptic type of Pell equations, which clearly have only finitely many integer solutions.

Take $r = 2, \delta = -57$ and $v + y = -7, u + x = 9$, that is, the matrix we consider is

$$\begin{bmatrix} 3 & 9 \\ -7 & -2 \end{bmatrix}; \ 1 + 4\delta = -227.$$

More precisely $\alpha = -1, \beta = 3, u = 0, v = -6, x = 9,$ and $y = -1$, i.e., the nil-clean decomposition

$$\begin{bmatrix} 3 & 9 \\ -7 & -2 \end{bmatrix} = \begin{bmatrix} 0 & 0 \\ -6 & 1 \end{bmatrix} + \begin{bmatrix} 3 & 9 \\ -1 & -3 \end{bmatrix}.$$

The (elliptic) Pell equation which corresponds to a unit with det $= 1$ is $X^2 + 227Y^2 = 15.371.300$ with $X = 3(227t + 777)$ (we shall not need Y).

Since $X = 227(3t + 10) + 61$ we deduce $X^2 = 227k + 89$ for a suitable integer k. However, since $15.371.300 = 67.715 \times 227 - 5$ from the Pell equation we obtain $X^2 = 227l - 5$ (for a suitable integer l) and so there are no integer solutions.

As for the equation which corresponds to det $= -1$, $X^2 + 227Y^2 = 16.478.700$ with $X = 3(227t+805)$. Analogously, $X = 227(3t+10)+145$ and $X^2 = 227p+141$ (for some integer p). Since from the Pell equation ($16.478.700 = 72.593 \times 227 + 89$) we obtain $X^2 = 227q+89$ (for an integer q), and again we have no integer solutions.

7.5.4   How the Example Was Found

A deceptive good news is that both equations (in Theorem 7.5.4) are solvable (over $\mathbb{Z}$): the first equation admits the solutions

$$X = \pm(v + y)(2r + 1)(2\delta + 3) \text{ and } Y = \pm(v + y)(2r + 1),$$

and the second equation admits the solutions:

$$X = \pm(v + y)(2r + 1)(2\delta - 1) \text{ and } Y = \pm(v + y)(2r + 1).$$

Therefore, the main problem which remains with respect to the solvability of the initial equations in s and t (γ is determined by s and t) is whether the linear systems above (in s and t) also have solutions (over $\mathbb{Z}$). Here is an analysis of this problem, just for the solutions given above.

For a unit with det $= 1$ we have four solutions:
for $+X = +(v + y)(2r + 1)(2\delta + 3)$ we obtain $t = 0$.
Then for $+Y = +(v + y)(2r + 1)$ we obtain

$$s = u + x + \frac{r^2 + 2r + 2}{v + y} \text{ and } \gamma = -1$$

and for $-Y = -(v + y)(2r + 1)$ we obtain

$$s = u + x + \frac{r^2 + 1}{v + y} \text{ and } \gamma = 0.$$

The corresponding clean decompositions are

$$
\begin{bmatrix} r+1 & u+x \\ v+y & -r \end{bmatrix} = \begin{bmatrix} 0 & u+x+\frac{r^2+2r+2}{v+y} \\ 0 & 1 \end{bmatrix} + \begin{bmatrix} r+1 & -\frac{r^2+2r+2}{v+y} \\ v+y & -r-1 \end{bmatrix}
$$

$$
= \begin{bmatrix} 1 & u+x+\frac{r^2+1}{v+y} \\ 0 & 0 \end{bmatrix} + \begin{bmatrix} r & -\frac{r^2+1}{v+y} \\ v+y & -r \end{bmatrix}.
$$

Notice that r^2+1 and $r^2+2r+2 = (r+1)^2+1$ are nonzero.
For $-X = -(v+y)(2r+1)(2\delta+3)$ we obtain

$$
t = (v+y)(1 + \frac{5}{1+4\delta})
$$

which is an integer if and only if $1+4\delta$ divides $5(v+y)$. However, this has to be continued with conditions on s.

For a unit with $\det = -1$ we also have four solutions:
for $+X = (v+y)(2r+1)(2\delta-1)$ we obtain $t = 0$. Then for $+Y = (v+y)(2r+1)$ we obtain

$$
s = u+x+\frac{r^2+2r}{v+y} \text{ and } \gamma = -1
$$

and for $-Y = -(v+y)(2r+1)$ we obtain

$$
s = u+x+\frac{r^2-1}{v+y} \text{ and } \gamma = 0.
$$

The corresponding clean decompositions are

$$
\begin{bmatrix} r+1 & u+x \\ v+y & -r \end{bmatrix} = \begin{bmatrix} 0 & u+x+\frac{r^2+2r}{v+y} \\ 0 & 1 \end{bmatrix} + \begin{bmatrix} r+1 & -\frac{r^2+2r}{v+y} \\ v+y & -r-1 \end{bmatrix}
$$

$$
= \begin{bmatrix} 1 & u+x+\frac{r^2-1}{v+y} \\ 0 & 0 \end{bmatrix} + \begin{bmatrix} r & -\frac{r^2-1}{v+y} \\ v+y & -r \end{bmatrix}.
$$

Notice that $r^2-1 = 0$ if and only if $r \in \{\pm1\}$ and $r^2+2r = 0$ if and only if $r \in \{0,2\}$.

For $-X = -(v+y)(2r+1)(2\delta-1)$ we obtain $t = (v+y)\left(1 + \frac{1}{1+4\delta}\right)$ which is an integer if and only if $1+4\delta$ divides $v+y$. Again, this has to be continued with conditions on s.

Generally the relations $\alpha^2 + \alpha + uv = 0$ and $\beta^2 + xy = 0$, do not imply that $v+y$ divides any of r^2+1, r^2-1, $r^2+2r = (r+1)^2-1$ or $r^2+2r+2 = (r+1)^2+1$ (recall that $r = \alpha + \beta$), nor that $1+4\delta$ divides $5(v+y)$ (and so does not divide $v+y$).

Searching for a counterexample, we need integers $\alpha, \beta, u, v, x, y$ such that $\alpha^2 + \alpha + uv = 0 = \beta^2 + xy$, and $v + y$ does not divide any of the numbers: $r^2 + 1, r^2 - 1$, $(r + 1)^2 - 1$ or $(r + 1)^2 + 1$.

Further, $1 + 4\delta$ should not divide $5(v + y)$ and, moreover, to cover the trivial idempotents, we add two other conditions.

Since idempotents and units are clean in any ring, we must add:

$$\det \begin{bmatrix} r + 1 & u + x \\ v + y & -r \end{bmatrix} \neq 0$$

(this way the nil-clean matrix is not idempotent, nor nilpotent) and

$$\det \begin{bmatrix} r + 1 & u + x \\ v + y & -r \end{bmatrix} \neq \pm 1,$$

(it is not a unit, and so nor unipotent), that is $\delta \notin \{0, \pm 1\}$.

Notice that if $r \in \{-2, -1, 0, 1\}$, then 0 appears among our two numbers $(r^2 - 1, (r + 1)^2 - 1)$ and the fraction is zero (i.e., an integer).

Since a matrix is nil-clean if and only if its transpose is nil-clean, we should have symmetric conditions on the corners $v + y$ and $u + x$, respectively. That is why, $u + x$ should not divide any of the numbers: $r^2 + 1, r^2 - 1, (r + 1)^2 - 1$, or $(r + 1)^2 + 1$, and further, $1 + 4\delta$ should not divide $5(u + x)$.

Further, we exclude clean decompositions which use an idempotent of type

$$\begin{bmatrix} 0 & 0 \\ k & 1 \end{bmatrix}.$$

In this case the unit (supposed with $\det = -1$) should be

$$\begin{bmatrix} r + 1 & u + x \\ (v + y) - k & -r - 1 \end{bmatrix}$$

and if its determinant equals -1 then $u + x$ divides $r^2 + r$. Since idempotent, nilpotent, unit and so nil-clean matrices have the same property when transposed, to the conditions above we add $u + x$ and $v + y$ do not divide $r^2 + r$.

By inspection, one can see that there are no selections of $u + x$ and $v + y$ less than ± 7 and ± 9, at least for $r \in \{2, 3, \ldots 10\}$, which satisfy all the above nondivisibilities.

Therefore $v + y = -7$, $u + x = 9$ is some kind of minimal selection. In order to keep numbers in the Pell equation as low as possible we choose $r = 2$ and so $\delta = -57$.

Indeed, our matrix verifies all *these exclusion conditions*: -7 and 9 do not divide any of $r^2 \pm 1 = 3, 5$, $(r + 1)^2 \pm 1 = 8, 10$ nor $r^2 + r = 6$; $1 + 4\delta = -227$ (prime number) does not divide $5 \times (-7) = -35$ nor $5 \times 9 = 45$, and $\delta \notin \{0, \pm 1\}$.

Remark. We found this example in terms of r, δ, $u + x$, and $v + y$. It was not obvious how to come back to the nil-clean decomposition, that is, to α, β, u, v, x, and y (indeed, this reduces to another elliptic Pell equation!). However, the following elementary argument showed more: there is only one solution, given by $(u, v) = (0, -6)$.

The system $\alpha + \beta = 2$, $u + x = 9$, $v + y = -7$, $\alpha^2 + \alpha + uv = 0 = \beta^2 + xy$ is equivalent to $(7u - 9v - 59)(7u - 9v - 54) + 25uv = 0$. Denote $t = 7u - 9v - 59$, hence $u = \frac{1}{7}(9v + t + 59)$. We obtain the equation

$$t(t + 5) + 25uv = 0.$$

Looking mod 5, it follows $t = 5k$, for some integer k. The equation simplifies to $k(k + 1) + uv = 0$. That is

$$k(k + 1) + \frac{1}{7}(9v + 5k + 59)v = 0.$$

Considering the last equation as a quadratic equation in k, we have

$$7k^2 + (5v + 7)k + 9v^2 + 59v = 0.$$

The discriminant of the last equation is

$$\Delta = (5v + 7)^2 - 28(9v^2 + 59v) = -227v^2 - 1582v + 49.$$

In order to have integer solutions for our last equation it is necessary $\Delta \geq 0$ and Δ to be a perfect square. The quadratic function

$$f(v) = -227v^2 - 1582v + 49$$

has the symmetry axis of the equation $v_{max} = -\dfrac{1582}{2 \cdot 227} < 0$, and $f(1) < 0$, hence there are no integers $v \geq 1$ such that $f(v) \geq 0$.

On the other hand, we have $f(-7) = 0$, giving $k = 2$, hence $t = 10$. Replacing in the equation (1) we obtain $6 - 7u = 0$, equation with no integer solution. Moreover, we have $f(v) < 0$ for all $v < -7$.

From the above remark, it follows that all possible integer solutions for v are -6, -5, -4, -3, -2, -1, 0. Checking all these possibilities we obtain $f(-6) = 37^2$ and then $k = -1$. We get $t = -5$, and equation (1) becomes $-6u = 0$, hence $u = 0$.

7.6   A Related Question

Since both unit-regular and nil-clean rings are clean, a natural question is whether these two classes are somehow related. First $\mathbb{Z}_3$ (more generally, any domain with at least 3 elements) is a unit-regular ring which is not nil-clean, and, $\mathbb{Z}_4$ (more generally, any nil clean ring with nontrivial Jacobson radical) is nil-clean but not unit-regular.

Finally, we give examples of nil-clean matrices in $\mathcal{M}_2(\mathbb{Z})$ which are not unit-regular, and unit-regular matrices which are not nil-clean.

Recall that the set of all the nontrivial nil-clean matrices in $\mathcal{M}_2(\mathbb{Z})$ is

$$\left\{ \begin{bmatrix} \alpha+\beta+1 & u+x \\ v+y & -\alpha-\beta \end{bmatrix} \mid \alpha,\beta,u,v,x,y \in \mathbb{Z}, \alpha^2+\alpha+uv = 0 = \beta^2+xy \right\},$$

and that the only nonzero unit-regular matrices with a zero second row are

$$\begin{bmatrix} a & b \\ 0 & 0 \end{bmatrix},$$

with (a,b) unimodular (i.e., a row whose entries generate the unit ideal) [see [103]].

Hence $\begin{bmatrix} 2 & 1 \\ 0 & 0 \end{bmatrix}$ is *unit-regular but not nil-clean* (nil-clean matrices have trace equal to 2,1 or 0; in the first case $\begin{bmatrix} 2 & 1 \\ 0 & 0 \end{bmatrix} - I_2$ is not nilpotent). Conversely, first notice that the nil-clean matrices with a zero second row are exactly the matrices $\begin{bmatrix} 1 & b \\ 0 & 0 \end{bmatrix}, b \in \mathbb{Z}$. Being idempotent, these are also unit-regular (so not suitable).

However, consider the nil-clean matrix (with our notations $\alpha = \beta = v = x = 0$, $u = 1, y = 2$)

$$A = \begin{bmatrix} 1 & 1 \\ 2 & 0 \end{bmatrix}.$$

Suppose A is unit-regular. Then, using an equivalent definition, $A = EU$ with $E = E^2$ and $U \in GL_2(\mathbb{Z})$. Since $\det A = -2 \neq \pm 1$, A is not a unit and so $E \neq I_2$. Hence $\det E = 0$ and from $\det A = \det E \cdot \det U$, we obtain a contradiction.

References

1. Acu, D.: Aritmetică şi teoria numerelor (Romanian). Universitatea "Lucian Blaga", Sibiu, Colecţia Facultăţii de Ştiinţe, Seria Matematică (1999)
2. Alder, H.L.: n and $n + 1$ consecutive integers with equal sums of squares. Am. Math. Mon. **69**, 282–285 (1962)
3. Alder, H.L., Alfred, U.: n and $n + k$ consecutive integers with equal sums of squares. Am. Math. Mon. **71**, 749–754 (1964)
4. Alfred, B.U.: n and $n + 1$ consecutive integers with equal sums of squares. Math. Mag. **35**, 155–164 (1962)
5. Alfred, B.U.: Consecutive integers whose sum of squares is a perfect square. Math. Mag. **37**, 19–32 (1964)
6. Ando, S.: A note on the polygonal numbers. Fibonacci Q. **19**(2), 180–183 (1981)
7. Andreescu, T.: Test Problem No. 3. USA Mathematical Olympiad Summer Program (2000)
8. Andreescu, T.: A note on the equation $(x + y + z)^2 = xyz$. Gen. Math. **10**(3–4), 17–22 (2002)
9. Andreescu, T.: Solutions to the Diophantine equation $(x + y + z + t)^2 = xyzt$. Studia Univ. "Babeş-Bolyai" Math. **48**(2), 3–7 (2003)
10. Andreescu, T.: On integer solutions to the equation $x^3 + y^3 + z^3 + t^3 = n$. Matematika Plus **3–4**, 19–20 (2002)
11. Andreescu, T.: Problem 38. Math. Horizons, **33** (1996)
12. Andreescu, T.: Note on the equation $x^2 + axy + y^2 = 1$. Seminarul de analiză matematică şi aplicaţii în teoria controlului, Universitatea de Vest din Timişoara. Preprint no. 141, pp. 1–6 (2002)
13. Andreescu, T., Andrica, D.: 360 Problems for Mathematical Contests. GIL Publishing House, Zalău (2003)
14. Andreescu, T., Andrica, D.: Ecuaţia lui Pell. Aplicaţii (Romanian). Caiete metodico-ştiinţifice, Matematică, Universitatea din Timişoara **15** (1984)
15. Andreescu, T., Andrica, D.: Ecuaţia lui Pell şi aplicaţii (Romanian). în "Teme şi probleme pentru pregătirea olimpiadelor de matematică" (T. Albu, col.), pp. 33–42. Piatra Neamţ (1984)
16. Andreescu, T., Andrica, D.: An Introduction to Diophantine Equations GIL Publishing House, Zalău (2002)
17. Andreescu, T., Andrica, D.: Note on a general negative Pell's equation. Octogon Math. Mag. **10**(2), 703–705 (2002)
18. Andreescu, T., Andrica, D.: Solvability and unsolvability of the Diophantine equation $ax^2 - by^2 = c$. Octogon Math. Mag. **10**(2), 706–709 (2002)
19. Andreescu, T., Andrica, D.: On a Diophantine equation and its ramifications, Coll. Math. J. **35**(1), 15–21 (2004)

© Springer Science+Business Media New York 2015
T. Andreescu, D. Andrica, *Quadratic Diophantine Equations*,
Developments in Mathematics 40, DOI 10.1007/978-0-387-54109-9

20. Andreescu, T., Andrica, D.: Equations with solution in terms of Fibonacci and Lucas sequences. An. Şt. Univ. Ovidius Constanţa Ser. Mat. **22**(3), 5–12 (2014)

21. Andreescu, T., Andrica, D.: Number Theory. Structures, Examples and Problems. Birkhäuser, Boston (2009)

22. Andreescu, T., Andrica, D., Cucurezeanu, I.: An Introduction to Diophantine Equations. Birkhäuser, Boston (2010)

23. Andreescu, T., Andrica, D., Feng, Z.: 104 Number Theory Problems. From the Training of the USA IMO Team. Birkhäuser, Boston (2007)

24. Andreescu, T., Feng, Z.: Mathematical Olympiads 1999–2000. Problems and Solutions from Around the World. Mathematical Association of America, Washington, DC (2001)

25. Andreescu, T., Gelca, R.: Mathematical Olympiad Challenges. Birkhäuser, Boston (2000)

26. Andreescu, T., Kedlaya, K.: Mathematical Contests 1995–1996. Mathematical Association of America, Washington, DC (1997)

27. Andrica, D., Buzeteanu, Ş.: On the reduction of a linear recurrence of rth order. Fibonacci Q. **23**(1), 81–84 (1985)

28. Andrica, D., Buzeţeanu, Ş.: Reducerea unei recurenţe liniare de ordinul doi şi consecinţe ale acesteia (Romanian). Gazeta Matematică seria A (2) **3**(3–4), 148–152 (1982)

29. Andrica, D., Călugăreanu, G.: A nil-clean 2×2 matrix over the integers which is not clean. J. Algebra Appl. **13**(6), 9 (2014)

30. Anning, N.: A cubic equation of Newton's. Am. Math. Mon. **33**, 211–212 (1926)

31. Banea, H.: Probleme de matematică traduse din revista sovietică KVANT (Romanian). Ed. Did. Ped. Bucureşti (1983)

32. Barbeau, E.J.: Pell's Equation. Springer, New York (2003)

33. Barnett, I.A.: A Diophantine equation characterizing the law of cosines. Am. Math. Mon. **62**, 251–252 (1955)

34. Barry, A.D.: L'équation Diophantine $x(x+1) = ky(y+1)$. Enseign. Math. (2) **25**(1–2), 23–31 (1979)

35. Bastida, J.R.: Quadratic properties of a linearly recurrent sequence. In: Proceedings of the 10th Southeastern Conference on Combinatorics, Graph Theory and Computing. Winnipeg (Utilitas Mathematica) (1979)

36. Bastida, J.R., DeLeon. M.R.: A quadratic property of certain linearly recurrent sequences. Fibonacci Q. **19**(2), 144–146 (1981)

37. Bencze, M.: PP. 1385. Octogon Math. Mag. **7**, 2:148 (1999)

38. Bennett, M.A., Walsh, P.G.: The Diophantine equation $b^2 X^4 - dY^2 = 1$. Proc. AMS **127**, 3481–3491 (1999)

39. Bitim, B.D., Özel, E.: On solutions of indefinite binary quadratic form. Int. J. Algebra **8**(18), 889–894 (2014)

40. Bloom, D.M.: Solution to problem 10238. Am. Math. Mon. **102**, 275–276 (1995)

41. Boju, V., Funar, L.: The Math Problems Notebook, Birkhauser, Boston, Basel, Berlin (2007)

42. Borevici, Z.I., Safarevici, I.P.: Teoria numerelor (Russian). Moscova (1982)

43. Bosma, W., Stevenhagen, P.: Density computations for real quadratic units. Math. Comp. **65**(215), 1327–1337 (1996)

44. Brown, E.: The class number and fundamental unit of $\mathbb{Q}(\sqrt{2p})$ for $p \equiv 1 \pmod{16}$ a prime. J. Number Theory **16**(1), 95–99 (1983)

45. Bumby, R.T.: The Diophantine equation $3x^4 - 2y^2 = 1$. Math. Scand. **21**, 144–148 (1967)

46. Buşneag, D., Boboc, F., Piciu, D.: Aritmetică şi teoria numerelor (Romanian). Editura Universitaria, Craiova (1999)

47. Buzeţeanu, Ş.: Grade de efectivitate în teoria calculului. Aspecte recursive, analitice şi combinatoriale (Romanian). Teză de doctorat, Universitatea din Bucureşti (1988)

48. Camillo, V.P., Khurana, D.: A characterization of unit-regular rings. Comm. Algebra **29**(5), 2293–2295 (2001)

49. Cavering, J.: L'irrationalité, dans les mathématiques grecques jusqu'a Euclid. Presses Universitaires du Septentrion, Paris (1998)

50. Carmichael, R.D.: The theory of numbers and Diophantine analysis. Dover, New York (1959)

51. Chao, K., Chi, S.: On the Diophantine equation $x^4 - Dy^2 = 1$, II. Chin. Ann. Math. Ser. A **1**, 83–88 (1980)
52. Chrystal, G.: Algebra, an Elementary Text-Book. Part II. Dover, New York (1961)
53. Cohen, E.: Theorie des numbers. Tome II, Paris (1924)
54. Cohen, H.: A Course in Computational Algebraic Number Theory. Springer, New York (1993)
55. Cohn, J.H.E.: The Diophantine equation $x^4 - Dy^2 = 1$ (II). Acta Arith. **78**, 401–403 (1997)
56. Coppel, W.W.: Number Theory, 2nd edn. Springer, New York (2009)
57. Cornacchia, G.: Su di un metodo per la risoluzione in numeri interi dell'equazione
$$\sum_{h=0}^{n} C_h x^{n-h} = P.$$ Giornale di Matematiche di Battaglini **46**, 33–90 (1908)
58. Cremona, J.E., Odoni, R.W.K.: Some density results for negative Pell equations; an application of graph theory. J. Lond. Math. Soc. (2) **39**(1), 16–28 (1989)
59. Davis, M., Matiyasevich, Yu., Robinson, J.: Diophantine equations: a positive aspect of a negative solution. Mathematical developments arising from Hilbert problems. Proc. Symp. Pure Math. **28**, 323–378 (1976). Am. Math. Soc., Providence
60. Davis, M., Putnam, H., Robinson, J.: The decision problem for exponential Diophantine equations. Ann. Math. **74**, 425–436 (1961)
61. DeLeon, M.J.: Pell's equation and Pell number triples. Fibonacci Q. **14**(5), 456–460 (1976)
62. Dickson, L.E.: Introduction to the theory of numbers. Dover, New York (1957)
63. Diesl, A.J.: Classes of strongly clean rings. PhD Thesis, Univ. of California, Berkeley (2006)
64. Diesl, A.J.: Nil clean rings. J. Algebra **383**, 197–211 (2013)
65. Dörrie, H.: 100 Great Problems of Elementary Mathematics. Their History and Solution. Dover, New York (1965)
66. Eckstein, Gh.: Fracţii continue (Romanian). RMT **1**, 17–35 (1986)
67. Edwards, H.M.: Fermat's Last Theorem. Springer, New York (1977)
68. Eggan, L.C., Eggan, P.C., Selfridge, J.L.: Polygonal products of polygonal numbers and the Pell equation. Fibonacci Q. **20**(1), 24–28 (1982)
69. Epstein, P.: J. Reine Angew. Math. **171**, 243–252 (1934)
70. Erdös, P.: On a Diophantine equation. J. Lond. Math. Soc. **26**, 176–178 (1951)
71. Faisant, A.: L'equation diophantienne du second degré. Hermann, Paris (1991)
72. Feit, W.: Some Diophantine equations of the form $x^2 - py^2 = z$. Proc. Am. Math. Soc. **129**(2), 623–625 (2000)
73. Finch, S.R.: Mathematical Constants. Encyclopedia of Mathematics and Its Applications, vol. 94. Cambridge University Press, Cambridge (2003)
74. Fowler, D.H.: The Mathematics of Plato's Academy: A New Reconstruction. Clarendon, Oxford (1987)
75. Ghelfond, A.O.: Rezolvarea ecuaţiilor în numere întregi (Romanian). Bucureşti (1954)
76. Gica, A.: Algorithms for the equation $x^2 - dy^2 = k$. Bull. Math. Soc. Sci. Math. Roum. Nouv. Sér **38**(3–4), 153–156 (1995)
77. Golomb, S.W.: Powerful numbers. Am.Math. Mon. **77**(8), 848–852 (1970)
78. Gopalan, M.A., Vidhyalakshmi, S., Kavitha, A.: Observations on $(x + y + z)^2 = xyz$. Int. J. Math. Sci. Appl. **3**(1), 59–62 (2013)
79. Gopalan, M.A., Vidhyalakshmi, S., Kavitha, A.: Integral solutions to the biquadratic equation with four unknowns $(x + y + z + w)^2 = xyzw + 1$. IOSR J. Math. **7**(4), 11–13 (2013)
80. Grosswald, E.: Boll. Un. Ital. (4) **1**, 382–392 (1968)
81. Grytczuk, Al., Luca, F., Wójtowicz, M.: The negative Pell equation and Pythagorean triples. Proc. Japan Acad. Ser. A Math. Sci. **76**(6), 91–94 (2000)
82. Guy, R.K.: Unsolved Problems in Number Theory. Springer, New York (1994)
83. Halici, S.: On some inequalities and Hankel matrices involving Pell, Pell-Lucas numbers. Math. Rep. **15**(65/1), 1–10 (2013)
84. Halter-Koch, F.: Diophantine equations of Pellian type. J. Number Theory **131**, 1597–1615 (2011)
85. Hansen, R.T.: Arithmetic of pentagonal numbers. Fibonacci Q. **8**(1), 83–87 (1970)

86. Hardy, K., Muskat, J.B., Williams, K.S.: A deterministic algorithm for solving $n = fu^2 + gv^2$ in coprime integers u and v. Math. Comp. **55**, 327–343 (1990)

87. Hardy, K., Muskat, J.B., Williams, K.S.: Solving $n = au^2 + buv + cv^2$ using the Euclidean algorithm. Utilitas Math. **38**, 225–236 (1990)

88. Hardy, G.H., Wright, E.M.: Theory of Numbers, 3rd edn. Oxford University Press, New York (1954)

89. Herschfeld, A.: Problem E1457. Am. Math. Mon. **68** 930–931 (1961)

90. Hoggatt, Jr., V.E.: Fibonacci and Lucas Numbers. The Fibonacci Association, Santa Clara (1969)

91. Hoggatt, V.E., Bicknell-Johnson, M.: A primer for the Fibonacci XVII: generalized Fibonacci numbers satisfying $u_{n+1}u_{n-1} - u_n^2 = \pm 1$. Fibonacci Q. **16**(2), 130–137 (1978)

92. Horadam, A.F.: Geometry of a generalized Simson's formula. Fibonacci Q. **20**(2), 164–168 (1982)

93. Hurwitz, A.: Lectures on Number Theory. Springer, New York (1986)

94. Hua, L.K., Wang, Y.: On Diophantine approximations and numerical integrations I, II. Sci. Sinica **13**, 1007–1009 (1964) and **13**, 1009–1010 (1964)

95. Hua, L.K.: Introduction to Number Theory. Springer, Berlin (1982)

96. Jones, J.P.: Diophantine representation of Fibonacci numbers. Fibonacci Q. **13**(1), 84–88 (1975)

97. Jones, J.P.: Diophantine representation of the Lucas numbers. Fibonacci Q. **14**(2), 134 (1976)

98. Jones, J.P.: Representation of solutions of Pell equations using Lucas sequences. Acta Academiae Pedagogicae Agriensis. Sectio Matematicae **30**, 75–86 (2003)

99. Kaplan, P.: Sur le 2-groupe des classes d'ideaux des corps quadratiques. J. Reine Angew. Math. **283/284**, 313–363 (1976)

100. Kedlaya, K.: What is an algebraic number?. Berkeley Math. Circle (December 2000)

101. Kenji, K.: On the Diophantine equation $x(x+1) = y(y+1)z^2$. Proc. Japan Acad. Ser. A Math. Sc. **72**(4), 91 (1996)

102. Keskin, R., Demiturk, N.: Solutions to some Diophantine equations using generalized Fibonacci and Lucas sequences. Ars Combinatoria **CXI**, 161–179 (2013)

103. Khurana, D., Lam, T.Y.: Clean matrices and unit-regular matrices. J. Algebra **280**, 683–698 (2004)

104. Knuth, D.E.: The Art of Computer Programming. Fundamental Algorithms, vol. I. Addison-Wesley, Boston (1975)

105. Koshy, T.: Fibonacci and Lucas Numbers with Applications. Wiley, New York (2001)

106. Koshy, T.: Pell and Pell-Lucas Numbers with Applications. Springer, New York (2015)

107. Krausz, T.: A note on equal values of polygonal numbers. Publ. Math. Debrecen **54**(3–4), 321–325 (1999)

108. Lagarias, J.C.: On the computational complexity of determining the solvability or unsolvability of the equation $X^2 - DY^2 = -1$. Trans. Am. Math. Soc. **260**(2), 485–508 (1980)

109. Lehmer, E.: On some special quartic reciprocity laws. Acta Arith. **21**, 367–377 (1972)

110. Lehner, J., Sheingorn, M.: Computing self-intersections of closed geodesics on finite-sheeted covers of the modular surface. Math. Comp. **44**(169), 233–240 (1985)

111. Lenstra, Jr., H.W.: Solving the Pell equation. Notices Am. Math. Soc. **49**(2), 182–192 (2002)

112. Leveque, W.J.: Topics in Number Theory, vol. 1. Addison-Wesley, New York (1956)

113. Li, X.-J.: On the Trace of Hecke Operators for Maass Forms. Number Theory (Ottawa, ON, 1996). CRM Proc. Lecture Notes, vol. 19, pp. 215–229. Am. Math. Soc., Providence (1999)

114. Li, Z.: The positive integer solutions of $ax^2 - by^2 = c$. Octogon Math. Mag. **10**(1), 348–349 (2002)

115. Lind, D.A.: The quadratic field $\mathbb{Q}(\sqrt{5})$ and certain Diophantine equations. Fibonacci Q. **6**(3), 91 (1968)

116. Ljunggren, W.: Einige Eigenschaften der Einheiten reel Quadratischer und rein- biquadratischen Zahlkorper. Skr. Norske Vid. Akad. Oslo (I) **1936**(12)

117. Ljunggren, W.: Zur Theorie der Gleichung $x^2 = Dy^4$. Avh. Norske Vid. Akad. Oslo (I) **1942**(5)

118. Ljunggren, W.: Some remarks on the Diophantine equations $x^2 - Dy^4 = 1$ and $x^4 - Dy^2 = 1$. J. Lond. Math. Soc. **41**, 542–544 (1966)
119. Ljunggren, W.: Some theorems on indeterminate equations of the form $(x^n - 1)/(x - 1) = y^q$. Norsk Mat. Tidsskr. **25**, 17–20 (1943)
120. Luca, F.: A generalization of the Schinzel-Sierpinski system of equations. Bull. Math. Soc. Sci. Math. Roumanie (N.S.) **41**(89/3), 181–195 (1998)
121. Luca, F., Togbe, A.: On the Diophantine equation $x^4 - q^4 = py^3$. Rocky Mountain J. Math. **40**(3), 995–108 (2010)
122. Luthar, R.S.: Problem E2606. Am. Math. Mon. **83**(8), 566 (1976)
123. Luthar, R.S.: Solution to problem E2606. Am. Math. Mon. **84**, 823–824 (1977)
124. Manin, Yu.I., Panchishkin, A.A.: Introduction to Modern Number Theory, 2nd edn. Springer, New York (2005)
125. Matthews, K.: Number Theory. Chelsea, New York (1961)
126. Matthews, K.: The Diophantine equation $x^2 - Dy^2 = N, D > 1$, in integers. Expos. Math. **18**, 323–331 (2000)
127. Matthews, K.: The Diophantine equation $ax^2 + bxy + cy^2 = N, D = b^2 - 4ac > 0$. J. Théor. Numbers Bordeaux **14**(1), 257–270 (2002)
128. Matthews, K.: Thue's theorem and the Diophantine equation $x^2 - Dy^2 = \pm N$. Math. Comp. **71**(239), 1281–1286 (2002)
129. Maohua, L.: A note on the Diophantine equation $x^{2p} - Dy^2 = 1$. Proc. AMS **107**(1), 27–34 (1989)
130. Matiyasevich, Yu.: Le dixième problème de Hilbert: que peut-on faire avec les équations diophantiennes? La Recherche de la Vérité, coll. L'écriture des Mathématiques, ACL - Les Éditions du Kangourou, pp. 281–305 (1999)
131. Matiyasevich, Yu.: Hilbert's Tenth Problem: Diophantine Equations from Algorithmic Point of View. Hilbert's Problems Today, 5th–7th April 2001. Pisa, Italy (2001)
132. Matiyasevich, Yu.: Hilbert's Tenth Problem. MIT, Cambridge (1993)
133. Matiyasevich, Yu.: Enumerable sets are Diophantine. Soviet Math. Doklady **11**, 354–358 (1970)
134. McDaniel, W.L.: Diophantine representation of Lucas sequences. Fibonacci Q. **33**(1), 59–63 (1995)
135. McLaughlin, R.: Sequences. some properties by Matrix methods. Math. Gaz. **64**(430), 281–282 (1980)
136. Megan, M.: Bazele analizei matematice (Romanian), vols I–III. Editura Eurobit, Timişoara (1995–1997)
137. Megan, M.: Analiză Matematică (Romanian), vol. I. Editura Mirton, Timişoara (1999)
138. Megan, M.: Analiză Matematică (Romanian), vol. II. Editura Mirton, Timişoara (2000)
139. Mignotte, M.: A note on the equation $x^2 - 1 = y^2(z^2 - 1)$. C. R. Math. Rep. Acad. Sci. Canada **13**(4), 157–160 (1991)
140. Mihailov, I.I.: On the equation $x(x + 1) = y(y + 1)z^2$. Gaz. Mat. Ser. A **78**, 28–30 (1973)
141. Mollin, R.A.: Quadratics. CRS, Boca Raton (1996)
142. Mollin, R.A.: Fundamental Number Theory and Applications. CRC, New York (1998)
143. Mollin, R.A.: A Simple criterion for solving of both $X^2 - DY^2 = c$ and $x^2 - Dy^2 = -c$. N.Y. J. Math. **7**, 87–97 (2001)
144. Mollin, R.A.: A continued fraction approach to the Diophantine equation $ax^2 - by^2 = \pm 1$. JP J. Algebra Number Theory Appl. **4**, 159–207 (2004)
145. Mollin, R.A., Cheng, K.: Palindromy and continuous ideals revisited. J. Number Theory **74**, 98–110 (1999)
146. Mollin, R.A., Srinivasan, A.: A note on the negative Pell equation. Int. J. Algebra **4**(19), 919–922 (2010)
147. Mollin, R.A., Srinivasan, A.: Central norms: applications to Pell's equation. Far East J. Math. Sci. **38**(2), 225–252 (2010)
148. Mordell, L.J.: On the integer solutions of the equation $x^2 + y^2 + z^2 + 2xyz = n$. J. Lond. Math. Soc. **28**, 500–510 (1953)

149. Mordell, L.J.: Corrigendum: integer solutions of the equation $x^2 + y^2 + z^2 + 2xyz = n$. J. Lond. Math. Soc. **32**, 383 (1957)
150. Mordell, L.J.: Diophantine equations. Academic, London (1969)
151. Nagell, T.: Introduction to Number Theory. Wiley, New York (1951)
152. Nagell, T.: On a special class of Diophantine equations of the second degree. Ark. Mat. **3**, 51–65 (1954)
153. Narkiewicz, W.: Elementary and Analytic Theory of Algebraic Numbers. PWN, Warsaw (1974)
154. Nathan, J.A.: The Irrationality of e^x for Nonzero Rational x. Am. Math. Mon. **105**(8), 762–763 (1998)
155. Newman, M.: A note on an equation related to the Pell equation. Am. Math. Mon. **84**(5), 365–366 (1977)
156. Nikonorov, Y.G., Rodionov, E.D.: Standard homogeneous Einstein manifolds and Diophantine equations. Arch. Math. (Brno) **32**(2), 123–136 (1996)
157. Nitaj, A.: Conséquences et aspects expérimentaux des conjectures abc et de Szpiro. Caen (1994)
158. Niven, I.: Quadratic Diophantine equations in the rational and quadratic fields. Trans. Am. Math. Soc. **52**, 1–11 (1942)
159. Niven, I., Zuckerman, H.S., Montgomery, H.L.: An Introduction to the Theory of Numbers, 5th edn. Wiley, New York (1991)
160. Nowicki, A.: Liczby kwadratowe (Polish). Olsztyn, Toruń (2012)
161. Nowicki, A.: Równanie Pella (Polish). Olsztyn, Toruń (2013)
162. O'Donnell, W.J.: Two Theorems Concerning Hexagonal Numbers. Fibonacci Q. **17**(1), 77–79 (1979)
163. O'Donnell, W.J.: A theorem concerning octogonal numbers. J. Recreational Math. **12**(4), 271–272 (1979/1980)
164. Olds, C.D.: Continued Fractions. New Mathematical Library, vol. 22. The Mathematical Association of America, Washington, DC (1963)
165. Olds, C.D., Lax, A., Davidoff, G.: The Geometry of Numbers. The Mathematical Association of America, Washington, DC (2000)
166. Ono, K.: Euler's concordant forms. Acta Arith. **78**(2), 101–126 (1996)
167. Oppenheim, A.: On the Diophantine equation $x^2 + y^2 + z^2 + 2xyz = 1$. Am. Math. Mon. **64**, 101–103 (1957)
168. Owings, Jr., J.C.: An elementary approach to Diophantine equations of the second degree. Duke Math. J. **37**, 261–273 (1970)
169. Owings, Jr., J.C.: Solution of the system $a^2 \equiv -1 \pmod{b}$, $b^2 \equiv -1 \pmod{a}$. Fibonacci Q. **25**(3), 245–249 (1987)
170. Panaitopol, L.: Câteva proprietăţi ale numerelor triunghiulare (Romanian). G.M.-A. **3**, 213–216 (2000)
171. Panaitopol, L., Gica, Al.: Probleme celebre de teoria numerelor (Romanian). Editura Universităţii din Bucureşti (1998)
172. Panaitopol, L., Gica, Al.: O introducere în aritmetică şi teoria numerelor (Romanian). Editura Universităţii din Bucureşti (2001)
173. Parker, W.V.: Problem 4511. Am. Math. Mon. **61**, 130–131 (1954)
174. Patz, W.: Über die Gleichung $X^2 - DY^2 = \pm c(2^{31} - 1)$. Bayer. Akad. Wiss. Math.-Natur. Kl. S.-B, 21–30 (1948)
175. Pavone, M.: A remark on a theorem of Serret. J. Number Theory **23**, 268–278 (1986)
176. van der Poorten, A.: A proof that Euler missed... Apéry's proof of the irrationality of $\zeta(3)$. Math. Intell. **1**, 195–203 (1979)
177. Pumplün, D.: Über die Klassenzahl und die Grundeinheit des reelquadratischen Zahlkorper. J. Reine Angew. Math. **230**, 167–210 (1968)
178. Putnam, H.: An unsolvable problem in number theory. J. Symb. Logic **25**, 220–232 (1960)
179. Rassias, M.Th.: Problem-Solving and Selected Topics in Number Theory. In the Spirit of the Mathematical Olympiads, Springer, New York, Dordrecht, Heidelberg, London, (2011)

180. Redei, L.: Bedingtes Artisches symbol mit Andwendung in der Klassenkörpentheorie. Acta Math. Sci. Hungar **4**, 1–29 (1953)

181. Redei, L.: Die 2-Ringklassengruppe des Quadratischen Zahlkörpus und die theorie der Pellschen Gleichung. Acta Math. Acad. Sci. Hungar. **4**, 31–87 (1953)

182. Rieger, G.J.: Über die Anzahl der als Summe von zwei Quadraten darstellnaren und in einer primen Restklasse gelegenen Zahlen unter einer positiven Schranke. II. J. Reine Angew. Math. **217**, 200–216 (1965)

183. Rockett, A.M.: Continued Fractions. World Scientific, River Edge (1992)

184. Rockett, A.M., Szusz, P.: Continued Fractions. World Scientific, River Edge (1992)

185. Rose, H.E.: A Course in Number Theory. Clarendon, Oxford (1988)

186. Rosen, D., Schmidt, T.A.: Hecke groups and continued fractions. Bull. Aust. Math. Soc. **46**(3), 459–474 (1992)

187. Roy, S.: What's the next Fibonacci numbers. Math. Gaz. **63**(429), 189–190 (1980)

188. Savin, D.: About a Diophantine equation. An. Şt. Univ. Ovidius Constanţa Ser. Mat. **17**(3), 241–250 (2009)

189. Savin, D.: On some Diophantine equations (I). An. Şt. Univ. Ovidius Constanţa Ser. Mat. **10**(1), 121–134 (2002)

190. Savin, D.: On some Diophantine equation $x^4 - q^4 = py^3$, in the special conditions. An. Şt. Univ. Ovidius Constanţa Ser. Mat. **12**(1), 81–90 (2004)

191. Savin, D., Ştefănescu, M.: Lecţii de aritmetică şi teoria numerelor (Romanian). Editura Matrix Rom, Bucureşti (2008)

192. Sándor, J.: Problema 7402. Gamma, Anul X (1–2) (1987)

193. Sándor, J.: Problema 22877. Mat. Lapok, Anul XCVII, **10** (1992)

194. Sándor, J., Berger, G.: Aufgabe 1025. Elem. Math. **45**(1), 28 (1990)

195. Scholz, A.: Über die Losbarkeit der Gleichung $t^2 - Du^2 = -4$. Math. Z. **39**, 93–111 (1934)

196. Serret, J.-A. (ed.): Oeuvres de Lagrange, vols I–XIV. Gauthiers-Villars, Paris (1877)

197. Seidenberg, A.: A new decision method for elementary algebra. Ann. Math. **60**(1954), 365–374

198. Sierpinski, W.: Elementary Theory of Numbers. Polski Academic Nauk, Warsaw (1964)

199. Sierpinski, W.: Ce ştim şi ce nu ştim despre numerele prime (Romanian). Editura Ştiinţifică, Bucureşti (1966)

200. Sierpinski, W.: 250 Problems in Elementary Number Theory. American Elsevier, New York/ PWN, Warszawa (1970)

201. Sierpinski, W.: Une théorème sur les nombres triangulairs. Elem. Math. **23**, 31–32 (1968)

202. Silvester, J.R.: Fibonacci properties by matrix methods. Math. Gaz. **63**(425), 188–191 (1979)

203. Skolem, T.A.: Diophantische Gleichungen. Ergeb. Math. Grezgeb., Band 5, Heft 4, Berlin (1938). Reprint Chelsea, New York (1950)

204. Smith, J.: Solvability Characterizations of Pell Like Equations. MS Thesis, Boise State University (2009)

205. Starker, E.P.: Problem E2151. Am. Math. Mon. **76**, 1140 (1969)

206. Stevenhagen, P.: The number of real quadratic fields having units of negative norm. Exp. Math. **2**, 121–136 (1993)

207. Stevenhagen, P.: A density conjecture for the negative Pell equation. In: Computational Algebra and Number Theory, Proc. 1992 Sydney Conf., pp. 187–200. Kluwer, Dordrecht (1995)

208. Sudan, G.: Geometrizarea fracţiilor continue. Editura Tehnică (1955)

209. Tallini, G.: Some new results on sets of type (m, n) in projective planes. J. Geom. **29**(2), 191–199 (1987)

210. Tano, F.: Sur quelques theorems de Dirichlet. J. Reine Angew. Math. **105**, 160–169 (1889)

211. Tarski, A.: A Decision Method for Elementary Algebra and Geometry. University of California Press, Berkeley (1951)

212. Tattersall, J.J.: Elementary Number Theory in Nine Chapters. Cambridge University Press, Cambridge (1999)

213. Titchmarsh, E.C.: The Theory of Riemann Zeta Function. Clarendon, Oxford (1951)

214. Utz, W.R.: Problem E3138. Am. Math. Mon. **96**(10), 928–932 (1989).

215. Vardi, I.: Archimedes' cattle problem. Am. Math. Mon. **105**(4), 305–319 (1998)

216. Venkatachalam, I.: Problem 4674. Am. Math. Mon. **63**, 126 (1956)

217. Vorobiev, N.N.: Fibonacci Numbers. Birkhäuser, Basel (2002)

218. ver der Waall, R.W.: On the Diophantine equations $x^2 + x + 1 = 3v^2$, $x^3 - 1 = 2y^2$, $x^2 + 1 = 2y^2$. Simon Stevin **46**, 39–51 (1972/1973)

219. Walker, D.T.: On the Diophantine equation $mx^2 - ny^2 = \pm 1$. Am. Math. Mon. **74**(5), 504–513 (1966)

220. Walker, D.T.: Consecutive integer pairs of powerful numbers and related Diophantine equations. Fibonacci Q. **14**, 111–116 (1976)

221. Walsh, G.: A note on a theorem of Ljunggren and the Diophantine equations $x^2 - kxy^2 + y^4 = 1, 4$. Arch. Math. **73**, 119–125 (1999)

222. Wang, Y.B.: On the Diophantine equation $(x^2 - 1)(y^2 - 1) = (z^2 - 1)^2$. Heilongjiang Daxue Ziran Kexue Xuebao (4), 84–85 (1989)

223. Wenchang, C.: Regular polygonal numbers and generalized Pell equations. Int. Math. Forum **2**(16), 781–802 (2007)

224. Whitford, E.E.: Pell Equation. Columbia University Press, New York (1912)

225. Williams, K.S.: On finding the solutions of $n = au^2 + buv + cv^2$ in integers u and v. Utilitas Math. **46**, 3–19 (1994)

226. Williams, H.C.: Solving the Pell's equation. In: Bennett, M.A., et al. (eds.) Proc. of Millennial Conf. on Number Theory, Urbana, 2000. A.K. Peters, Boston (2002)

227. Wojtacha, J.: On integral solution of a Diophantine equation. Fasc. Math. **8**, 105–108 (1074/1975)

228. Wu, H., Le, M.: A note on the Diophantine equation $(x^2 - 1)(y^2 - 1) = (z^2 - 1)^2$. Colloq. Math. **71**(1), 133–136 (1996)

229. Wulczyn, G.: Problem E2158. Am. Math. Mon. **76**, 1144–1146 (1969)

230. Zeckendorf, E.: Représentation des nombres naturels par une somme de nombres de Fibonacci ou des nombres de Lucas. Bull. Soc. R. Sci. Liège **41**, 179–182 (1972)

231. Zhenfu, C.: On the Diophantine equations $x^2 + 1 = 2y^2$ and $x^2 - 1 = 2Dz^2$. J. Math. (Wuhan) **3**, 227–235 (1983)

232. Zhenfu, C.: On the Diophantine equation $x^{2n} - Dy^2 = 1$. Proc. AMS **98**(1), 11–16 (1986)

Index

© Springer Science+Business Media New York 2015
T. Andreescu, D. Andrica, *Quadratic Diophantine Equations*,
Developments in Mathematics 40, DOI 10.1007/978-0-387-54109-9

Printed in the United States
By Bookmasters